Central Heating Wet a

Fourth Edition

Technical Editor: Chris Long

Technical Illustrator: Chris Long

Original Technical Illustrator: Graham Elkins

A CIP Catalogue record for this manual is available from the British Library. ©2012.

Notice

Acknowledgement

CORGI*direct* gratefully acknowledges the use in this manual of reference material published by the British Standards Institution (BSi) and Building Regulations (England and Wales).

Published by

CORGI*direct*

CORGI*direct*

Telephone: 0800 915 0490
Website: www.corgi-direct.com

First Edition:	April 2000
Second Edition:	October 2003
Third Edition	January 2007
Reprinted:	October 2007
Third Edition – Revised:	June 2009
Fourth Edition:	February 2011
Reprinted:	June 2011
Reprinted:	June 2012

Printed by: Blackmore Ltd

ISBN: 978-1-907723–02–5

Contents

Contents

Contents

Contents

Figures

Tables

What this guidance covers

This Part gives guidance for installation, service and maintenance of central heating boilers, combination boilers, condensing boilers and combined gas fire/back boiler/back circulators that are 'CE' or British Standard (BS) kite marked.

The guidance also applies for a used or second-hand appliance that doesn't carry the CE or BS kite-mark, but has a data plate confirming the appliance is suitable for the gas type and pressure i.e. Natural gas at an appliance inlet pressure of 20mbar and Liquefied Petroleum Gas (LPG) Propane and Butane where the supply regulator has been set to provide an operating pressure of 37mbar and 28mbar respectively.

Note: Read this Part together with the relevant part(s) of the current Gas Installer Manual Series – Essential Gas Safety – Domestic. See Part 14 – CORGI*direct* Publications.

The competence you need to carry out work

- you must not carry out any work (see **Part 13 – Definitions**) in relation to gas appliances and other gas fittings or gas storage vessels covered by this manual unless you are competent to do so
- when any work is carried out, gas installing businesses must be registered with Gas Safe Register® and their gas operatives must hold valid certificates of competence for each work activity they wish to undertake.
- the certificate must be issued under the Nationally Accredited Certification Scheme (ACS) for individual gas fitting operatives (see Note) - or through the ACS aligned Scottish/National Vocational Qualification (S/NVQ) route
- no employer, member of the public or other responsible person should knowingly employ any gas operative who cannot comply with these requirements
- you must install all appliances and other gas fittings in accordance with the Gas Safety (Installation and Use) Regulations, British Standards, Building Regulations, Regulations for Electrical Installations or those Regulations appropriate to the geographical region in which they are to be installed
- you must also install appliances to comply both with the Water Supply (Water Fittings) Regulations – and manufacturers' installation instructions

Note: Gas Safe Register are facilitating an industry wide review of the current competency framework, which may bring about changes to the ACS system in the future. Information on any proposed changes will be communicated as and when they happen.

This manual uses data and information from various sources, which CORGI*direct* believes reflects current custom and practice within the sector.

The heating system – a background history

Gas central heating boilers covered here mostly trace their origin to high thermal capacity (water content) cast iron floor-mounted solid fuel boilers.

In these, gravity instigated the movement of hot water around the circuit – a differential in density between hot water flow (lighter) and cool water return (heavier). This gravity circulation was generally facilitated via large capacity single cast iron/steel pipe circuits, installed on the internal perimeter walls of a building. Heating the building relied on the pipework acting as a heat emitter.

The later high thermal capacity radiators still relied on gravity circulation and pipework needed a rise from or fall back to the boiler. Water circulation was still fairly slow. Due to pipe sizes and high installation costs, central heating was generally confined to factories, churches and large country houses.

Temperature control of the hot water storage vessel (cylinder) was generally achieved by using a boiler thermostat - a basic control method refined by fitting a manually adjustable thermostatic control valve in the return pipework from the storage vessel.

Due to lack of maintenance, these valves often seized (when closed or semi-closed) - resulting in poor circulation. Alternatively, an electrically operated valve (controlled by a cylinder thermostat) was sometimes fitted in the flow or return pipework to the storage vessel.

A change for the better in the 1960s

In the 1960s gas central heating was introduced to the domestic market.

Most boilers were of the floor standing cast iron types, but they kept the high thermal capacity (13-14 litres water content) derivatives of solid fuel boilers. Circulation to radiators was improved by an accelerator (circulating pump) introduced into the small-bore central heating pipework system.

Circulation was quite often via a small-bore (15mm equivalent size) copper single pipe system – see **Single pipe systems: problems that exist and how to upgrade** in this Part.

Water circulation to the domestic hot water storage vessel was still driven by gravity – even an innovative back boiler introduced at the time kept the option of gravity circulation to the storage vessel.

How the feed and expansion cistern worked

The main safety feature was another solid fuel method/practice. If the boiler thermostat failed, water would boil, expand and discharge through an open vent pipe into the feed and expansion cistern (see Note), then return to the boiler via a cold feed supply pipe

This cycle continued until the user became aware of the noise generated by the boiling water, at which point they could hastily shut the system down

Note: Expansion cisterns must be installed correctly to ensure safety. See 'Building regulation requirements – CWSC – safety considerations' in this Part and Part 8 – 'General installation details – Wet central heating' of this manual.

Single pipe systems: problems that exist and how to upgrade

Single pipe systems are no longer installed, but many still exist. Some problems are described here, with advice on how to upgrade such systems.

A single pipe system is a simple ring circuit. A central heating system may have one circuit serving the whole house (see Figure 1.1), or it may have multiple circuits (see Figure 1.2), with each circuit serving an individual part of a dwelling.

For example, a three bedroom semi-detached property may have two circuits, one serving the upstairs radiators, the other serving the downstairs radiators.

Although the main circulation through the pipework is pumped, the water circulates through the radiators mainly by gravity.

As the water is pumped around the circuit there is a gradual pressure loss between the circulating pump outlet and inlet.

Figure 1.1 Single pipe ring circuit

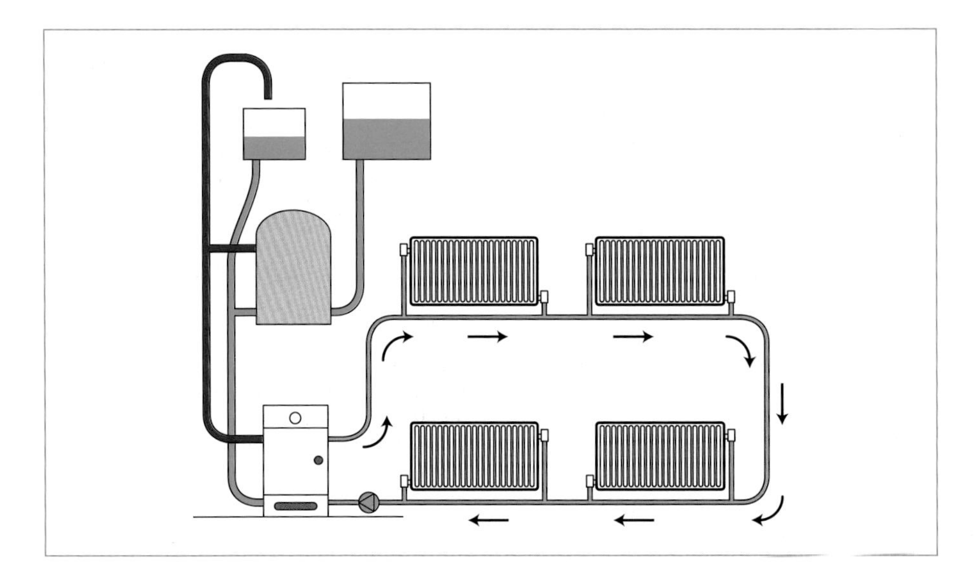

Figure 1.2 Single pipe circuit with two ring circuits

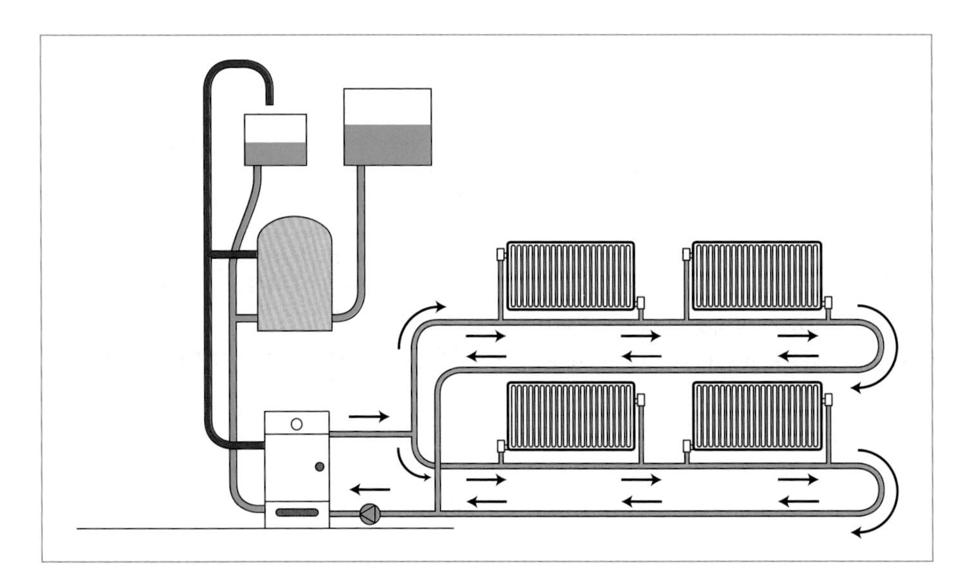

So there is then a differential in pressure between the tee pieces serving each radiator – and this also assists circulation to each radiator.

Because circulation to each radiator was by gravity, the pipe circuit had to be installed close to the radiator, with the radiator inlet and outlet connecting pipe tee pieces installed close to the radiator valves.

If this was not done, there was either poor circulation, or even no circulation at all through the radiator.

As circulation to a radiator is dependent on gravity circulation and a slight differential in water pressure between the inlet and outlet connections of the radiator, you can see that the greater the distance between the tee pieces, the greater the differential in pressure will be - assisting the flow through the radiator. So there was clearly little force available to aid circulation to a radiator on a single pipe system.

- so when you re-site radiators to redirect the main pipe circuit, you must redirect as close as possible to the new radiator position
- if you fail to do this, it will almost certainly result in poor circulation or no circulation to the radiator

In single pipe systems the domestic hot water storage vessel flow and return pipework often relies on gravity circulation only (see **Hot water storage vessels and heat recovery** in this Part).

- as well as allowing for a loss of pressure through the system pipework, you must allow for the loss of temperature through the system

Water flowing through each radiator loses heat into the room before subsequently flowing back into the same system pipework and onto the next radiator. This reduces the water temperature to the next and subsequent radiator(s) on the system.

- you must allow for this by sizing each radiator based on the projected water flow temperature available to the radiators on a single pipe system. In most cases this will mean larger radiators

Upgrading single pipe systems with thermostatic radiator valves (TRV)

- when you upgrade these systems with thermostatic radiator valves, only use valves suitable for a single-pipe system
- TRVs designed for fully pumped systems are not suitable - as the reduced valve area creates a high resistance to water flow. This often prevents circulation - or certainly reduces it
- another problem you will find is where TRVs are mistakenly fitted to the bottom radiator connections
- warm convection air currents from the hot main central heating circuit below the radiator often cause premature closure of the valve, with the valve remaining closed during the heating period. Mysteriously, the 'fault' corrects itself on initial start-up of the system (with everything cool), but followed shortly afterwards by heat from the main central heating circuit - causing premature closure again

Note: It can be seen that single pipe systems are not particularly effective or efficient and where compared to today's methods of installation, are an antiquated design. Where upgrading single pipe systems, every effort should be made to adopt modern design/installation standards.

Figure 1.3 A basic two pipe central heating system

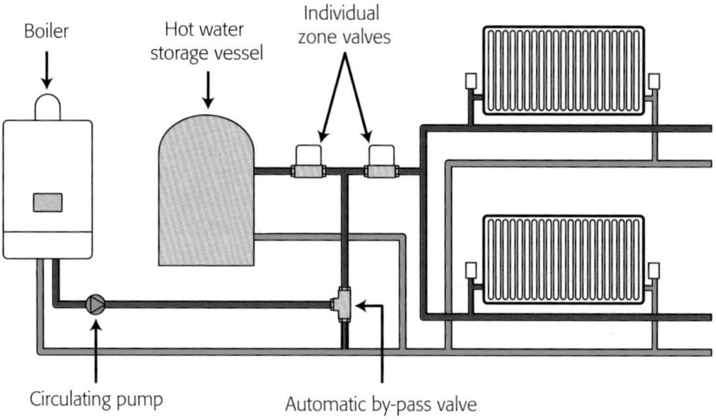

Two pipe systems – modern systems designed around these

Early two pipe systems were generally a refinement of single pipe systems, and still relied on gravity fed hot water to the hot water storage vessel.

Modern systems will almost certainly be designed around a two-pipe system, heating all rooms simultaneously to full comfort temperatures, see Figure 1.3 (see also **Comfort conditions – the basic requirements** in this Part).

The system will almost certainly incorporate:

- a programmer/timer to control heating and hot water periods
- thermostats to control the temperatures of both domestic hot water (DHW) and central heating (CH)
- separate 'zone' control of the heating circuit (see **Building regulation requirements** in this Part); and
- electrical interlock of the boiler and its system

The system will be fully pumped with individual flow and return pipes to each radiator or manifolds, in the case of underfloor heating circuits. The boiler may be either floor-standing, wall-mounted or a fire/back boiler serving the needs of both heating and hot water demands.

- alternatively, you may install a combination boiler serving the needs of both the heating and hot water in a combined unit
- you normally install combination boilers onto 'sealed systems' (see **Part 4 – Combination boilers – Sealed system** for further guidance)
- combination boilers have evolved over the years to become better specified, easier to install and to achieve even greater efficiencies – operating as most of them do in condensing mode.

Non-condensing models are available, but they are slowly disappearing from the market place (see **Part 5 – Condensing boilers** for further guidance)

Arguably, most systems are still installed utilising the open vent design, with the traditional separate open vent pipe that is terminated over a feed and expansion cistern, sited at the highest point in the installation e.g. the roof space.

The cistern will be supplied with water from the cold water main and have a separate cold feed supplying water to the circulating pipework.

- do not fit a valve to the cold feed pipe. If closed, this may prevent expansion of the water during heating
- modern boilers are equipped with an integral overheat thermostat – or this may be an optional extra from the boiler manufacturer
- the overheat thermostat is generally set to operate at a temperature of 95°C to prevent the system water from boiling

In some circumstances, you can install a combined cold feed/open vent arrangement - especially in the case of a low head installation.

This is where the boiler is sited 'high up' in the building, generally less than 1m head (the level of the water in the feed and expansion cistern above the boiler). Then, where the appliance incorporates an overheat thermostat and manufacturer's instructions allow, the open vent and cold feed may be combined in a single pipe.

In this case, the system does not need to have a separate open vent - only a suitably sized cold feed and expansion cistern to accommodate the hot expanded water.

- when you use this method, the boiler manufacturer's instructions often stipulate that the combined cold feed and open vent pipe must not be less than 20mm internal diameter
- there should be no valves or components - other than full-bore pipe fittings between the boiler and the feed and expansion cistern

Boilers fitted with this overheat cut-off device are often used as a 'package' with other controls on sealed systems.

Heating and hot water for today's needs

Boilers are now reduced in size and thermal capacity – and radiators and hot water storage vessels have also changed. Boilers today usually serve smaller thermal capacity radiators and fully pumped fast recovery hot water storage vessels. The storage vessel may be remote from the boiler, or it may be combined with the boiler within the boiler case/chassis.

Designing wet central heating systems can be complex

Because of the varying complexities and need for correct design, CORGI*direct* offer a Wet Central Heating System Design Guide to help you when dealing with wet central heating systems.

This and more is available from CORGI*direct*, visit www.corgi-direct.com.

The information that follows is only general guidance.

Hot water storage vessels and heat recovery

On older gravity systems, using a 15kW output boiler and standard hot water storage vessel (900mm x 450mm), gravity circulation can take up to 90 minutes to heat the water (this figure varies depending on how good or bad circulation is to the vessel).

In contrast, a standard hot water storage vessel with pumped primaries will deliver approximately 175 litres of hot water in 40 minutes.

A fast recovery vessel delivers 340 litres of hot water in the same 40 minute period - and takes up significantly less space in the airing cupboard.

Fast recovery storage vessel

A modern, fully pumped fast recovery hot water storage vessel that incorporates a multi-strand heat exchanger and serves the average 3 bedroom semi-detached household, is likely to have a storage capacity of 45 litres, stand 600mm high and have a diameter of 350mm.

This is a lot smaller than the average gravity-fed hot water storage vessel of yesteryear. Provided an appropriate boiler and controls are fitted (see **Approved Document G – points to note** in this Part), recovery times for these modern vessels are typically 8 minutes from cold from a 15kW output boiler.

- this size of vessel is ideally suited for families who prefer to shower – and can also supply a standard length (1700mm) bath
- 60 litre and larger options are available for the larger than average household - or where there's an exceptional demand for hot water

Comfort conditions – the basic requirements

The human body is probably at its most comfortable when doing the least amount of work. A healthy body should maintain its temperature at a steady 36.9°C.

If the body starts to get cold, the pores close and shivering produces rapid movement to try to raise the body temperature to normal. If the body becomes too hot, pores open and sweating occurs. As the moisture evaporates, it takes heat from the body and lowers the body temperature.

The body is comfortable when, during any activity, the heat it produces exactly matches the amount of heat being lost. When the body has no work to do its temperature is kept steady.

To feel fresh there needs to be air movement in a room. Excessive movement or draughts make us feel cold, even if air temperature is still quite high. In summer, using a fan helps us keep cool - though it does not actually reduce air temperature. What it does do is help to increase the rate at which perspiration evaporates. This increases the rate at which heat is taken from the body.

How to design comfort conditions in a dwelling

- design full central heating to maintain comfort conditions with an outside temperature of -1°C (see Note)

Note: The use of -1°C is a typical outside air temperature figure. The nature of the dwelling under consideration – its method and type of construction, its location (exposed sites, geographical location within the UK), etc – may require differing design figures.

- however, to reach and maintain the temperatures in Table 1.1 when the outside temperature is -1°C or less, you may need to set the boiler thermostat control to its maximum setting and override the heating time controls. This includes setting programmers to 24-hour operation and setting room or radiator thermostats to their maximum settings
- similarly, during a cold period, if a system has been off overnight or during the day, rooms might not be able to reach the temperatures shown in Table 1.1 within a short period of turning on the heating system, unless the room has an additional heat source e.g. a gas fire

  the fabric of a building or room left unheated will cool. Consequently, it will absorb most of the heat being produced once the heating system is turned on. This may take several hours in traditional built dwellings with minimal or no insulation installed

Table 1.1 Temperature and air changes on which heat loss calculations are based

Room	Room temperature °C[a]	Air changes per hour[b]
Living room	21	1.5
Dining room	21	1.5
Bedsitting room	21	1.5
Bedroom[c]	18	1
Hall/landing	18	1.5
Bathroom	22	2
Kitchen	18	2
Toilet	18	2

[a] These are the temperatures recommended for whole house central heating and for individual rooms with part central heating. Where open-flue appliances are installed the number of air changes should be increased – BS EN 12831 provides further detail.

[b] Figures in this Table are based on individual rooms that are isolated from each other e.g. with interconnecting doors closed and with the air changes so described. A room that contains a solid fuel chimney and fire grate that is open to the room may be subject to a greater number of air changes – especially if the solid fuel fire or decorative fuel effect gas appliance is alight – in which case it may be difficult to reach the desired temperatures indicated in this Table.

[c] When bedrooms are used as bedsitting rooms, or for studying purposes, a higher temperature may be required.

Where higher design temperatures are required – for the elderly or infirmed for example – all rooms should be designed with a temperature of 23°C.

- BS EN 12831: 2003 Heating systems in buildings – Method for calculation of the design heat load, recommends that 'comfortable' temperatures for activities and air changes in a dwelling should be as shown in Table 1.1

Note: Both BS EN 12831 and BS EN 12828: 2003 'Heating systems in buildings. Design for water-based heating systems' – replace the previous installation standard BS 5449: 1990 'Specification for forced circulation hot water central heating systems for domestic premises', which is withdrawn.

Building regulation requirements

The Building Regulations, specifically the Approved Documents (ADs) have gone through significant revisions in recent years; 1st April 2002 saw the first of these revisions and more recently, 2005/6 and 1st October 2010.

With regards to this manuals scope, the following ADs to the Building Regulations (England and Wales) are of importance:

- Approved Document G (ADG) covers sanitation, hot water safety and water efficiency
- Approved Document J (ADJ) covers combustion appliances and fuel storage systems
- Approved Document L1A (ADL1A) covers the conservation of fuel and power in new dwellings.
- Approved Document L1B (ADL1B) covers the conservation of fuel and power in existing dwellings

Note: There are two additional parts to ADL, ADL2A and ADL2B covering buildings other than dwellings, which are not discussed in this manual.

- Approved Document P (ADP) covers the design and installation of electrical installations

Note: Although the above ADs refer directly to Building Regulations for England and Wales, their various requirements including what's covered by this manual are mirrored within the Technical Handbooks for Scotland and Technical Booklets for Northern Ireland – albeit under differently labelled guidance documents – and therefore in the contexts of this manual are relevant to these regions.

Approve Document G – points to note

- ADG has revised and enhanced the requirements of G3 'How water supply and systems' to include safety provisions to all types of hot water systems and to the prevention of scalding

In particular –

- ADG3(1) has a new requirement to provide hot water to baths, bidets, showers, washbasins and sinks
- ADG3(2) extension of previous requirement to ensure safe operation to all types of hot water systems i.e. including controls and safety devices to limit temperature to a maximum of 100°C
- ADG3(3) extension of previous requirement to include all hot water systems, i.e. primary thermal stores
- ADG3(4) a new requirement to prevent scalding by the installation of protective devices that limit the temperature of hot water supplied to baths e.g. in-line blending valves

Note: ADG has been significantly updated to cover areas not discussed within this manual i.e. supply of wholesome water, supplies to sanitary conveniences, water efficiency in dwellings, etc. – further guidance on these aspects needs to be sought from the ADG.

Controlling hot water temperature:

Where the hot water is supplied from a hot water storage vessel, make sure the stored water temperature is controlled by a cylinder thermostat. Additionally, the following safety device(s) are required –

Vented hot water storage – directly heated

- In addition to a vent pipe which connects to the top of the storage vessel and terminates over and above the cold water storage cistern (CWSC) – see also **CWSC – safety considerations** in his Part –
  a non-self-resetting energy cut out needs to be installed (see **Non-self-resetting energy cut-outs** in this Part), which will prevent the stored water from over heating

Vented hot water storage – indirectly heated

- In addition to a vent pipe which connects to the top of the storage vessel and terminates over and above the CWSC, an overheat cut-out needs to be installed which will prevent the stored water from exceeding 100°C; or
- a safety device such as a temperature relief valve or a combined temperature and pressure relief valve needs to be installed

Unvented hot water storage vessel

- two independent devices need to be installed in addition to any thermostat used to control the stored water temperature, for example a non-self-resetting energy cut-out to disconnect the heat source and a temperature or a combined temperature and pressure relief valve to safely discharge the over heated water

Note: For unvented hot water storage vessels up to 500ltrs and 45kW and which are heated indirectly by a boiler, the energy cut-out may be on the boiler.

Non-self-resetting energy cut-outs

Only use non-self-resetting energy cut-outs which will instantly disconnect any heat source to the storage vessel.

Where the cut-out indirectly acts to disconnect the heat source – controls a motorised valve which shuts off the flow to the vessel – they need to comply with BS EN 60730-2-9: 2010 'Automatic electrical controls for household and similar use. Particular requirements for temperature sensing control' – or BS EN 257: 2010 'Mechanical thermostats for gas burning appliances'.

Where the heat source is via an immersion heater – which is outside the scope of this manual – this needs to comply with BS EN 60335-2-73: 2003 + A2: 2009 'Household and similar electrical appliances. Safety. Particular requirements for fixed immersion heaters'.

Hot water storage system: requirements to follow

Vented systems:

- with vented copper hot water storage vessels, ensure that the installation of products conforming to the relevant standards (i.e. BS 1566: 2002 + A1: 2011 'Copper indirect cylinders for domestic purposes. Open vented copper cylinders. Requirements and test methods' and BS 3198: 1981 'Specification for copper hot water storage combination units for domestic purpose')

Unvented systems:

- with unvented hot water storage systems, ensure that the installation of products complying with BS EN 12897: 2006 'Water supply. Specification for indirectly heated unvented (closed) storage water heaters' and/or certified by an accredited body i.e. the British Board of Agrément (BBA), or Water Research Council (WRc) comply with regulations

Vessel labelling

Regardless of the hot water storage vessel type, a suitably worded notice produced by the manufacturer should be displayed on the vessel, which provides the following information:

- type of vessel (vented, un-vented, etc)
- nominal capacity in litres
- standing heat loss in kWh/day
- heat exchanger performance in kW
- reference to the compliance product standard e.g. BS 1566 and logo's of accredited bodies, as required

Hard water areas

Where boilers, particularly combination boilers are installed in hard water areas, i.e. those where the mains total water hardness exceeds 200ppm and where required by the appliance manufacturer, additional provision needs to be made to treat the feed water to limit the formation of lime scale.

Suitable treatments include chemical limescale inhibitors, combined corrosion/limescale inhibitors, polyphosphate dosing, electrolytic scale reducers or water softeners.

Further guidance should be sought from the appliance manufacturer as to the preferred method of appliance/system protection.

Insulation

Pipework connected to hot water storage vessels, including vent pipes, needs to be insulated to a minimum of 1m from their point of connection.

Similarly, any pipework routed through unheated spaces needs to be insulated throughout their entire length, where the building structure allows.

CWSC – safety considerations

ADG reiterates the requirements for CWSCs to meet the relevant standards, for example BS 417-2: 1987 'Specification for galvanized low carbon steel cisterns, cistern lids, tanks and cylinders. Metric units' or BS 4213: 2004 'Cisterns for domestic use. Cold water storage and combined feed and expansion (thermoplastic) cisterns up to 500 l. Specification'.

All CWSC need to be supported on a flat, level, rigid platform, which is capable of safely withstanding the weight of the CWSC when filled with water to the rim.

The platform needs to support the whole base of the CWSC and extend outward to a minimum of 150mm on all sides of the CWSC.

Important: Cases have been recorded whereby CWSCs which have been inadequately supported and whilst being subjected to a fault condition – over heated water discharged in to the CWSC – have collapsed leading to dangerously hot water being ejected over persons below. This has tragically lead to the loss of life – ensure the full requirements for CWSCs are adhered to!

Approved Document J (ADJ) – points to note

- ADJ imposes greater restrictions on where you site chimney terminations in relation to openings into buildings and boundaries of properties. This brings it into line with the requirements of BS 5440-1: 2008 'Flueing and ventilation for gas appliances of rated input not exceeding 70 kW net (1st, 2nd and 3rd family gases). Specification for installation of gas appliances to chimneys and for maintenance of chimneys'
- new guidance is included on access for visual inspection of chimneys concealed within voids, i.e. those typically found in multi-occupancy dwellings (flats) where room-sealed chimneys are run within ceiling voids

Note: See Part 8 – General installation details – Wet central heating – Chimneys in voids – for further guidance

- ADJ also requires that notice plates are fitted for categories of chimney systems – to indicate their method of construction and suitability for use
- an increase in ventilation requirements for open-flued gas appliances installed in to very air tight houses (air permeability less than or equal to $5m^3/h/m^2$) – adventitious ventilation not sufficient in these properties
- further guidance is also included for solid fuel, liquid bio-fuel and oil (including blends of mineral oil) – not covered in this manual

Approved Document L (ADL1A and ADL1B) – points to note

- ADL specifies minimum efficiency standards for boilers and hot water storage vessels - for both when you install them in new dwellings and also when you replace them in existing dwellings
- you also need to fully commission heating and hot water systems and, where appropriate, provide the gas user with operating and maintenance instructions
- ADL is further explained/supported by the 'Domestic Building Services Compliance Guide' (DBSCG), which provides practical information to those having to comply with the requirements of ADL1A and ADL1B

Note: Similar guidance is also available for buildings other than dwellings (ADL2A and ADL2B), entitled 'Non-Domestic Building Services Compliance Guide'.

The requirements of ADL and therefore, the Building Regulations are closely linked with SAP (the Government's Standard Assessment Procedure for Energy Rating of Dwellings), SEDBUK (Seasonal Efficiency of Domestic Boilers in the UK) and the Code for Sustainable Homes.

ADL focuses on the conservation of fuel and power in all types of buildings and sets out requirements for dwellings and other buildings, outlining construction techniques, lighting, insulation and heating, to promote the most economic use of fossil fuels.

There are four main aspects to ADL:

1. It is necessary to reduce heat escaping from the fabric of the building.
2. Roofs, walls, windows, doors and floors should have an adequate resistance to heat loss.
3. Hot water pipes, hot air ducts and hot water storage vessels should also limit any heat loss.

4. Space heating and hot water systems must be energy efficient and have adequate controls provided to control appliances/heating and hot water systems - to avoid inefficient usage and waste.

- provide customers with the right information to help them operate and maintain heating and hot water systems economically

Building Regulations continue to evolve, bringing evermore stringent requirements given that buildings contribute to a high proportion of Carbon Dioxide (CO_2) emissions and the government has targets for reducing CO_2 levels.

- for compliance in dwellings, various energy efficiency-rating methods are used to show reasonable provision has been made for the conservation of fuel
- both new and modified dwellings that are within the regulations must now be subject to an Energy Rating: SAP

Standard Assessment Procedure (SAP)

SAP is the government's methodology of determining dwelling efficiencies, under typical installed conditions within Great Britain.

These conditions include:

- our climate
- housing conditions
- occupancy patterns; and
- controls (heating, etc)

For the control element, SAP 2009 (which is the latest version and which is more accurate than the previous SAP 2005 version), draws on another method of assessing efficiency that being Seasonal Efficiency of Domestic Boilers in the United Kingdom (SEDBUK) – see **Seasonal efficiency of domestic boilers in the United Kingdom (SEDBUK)** in this Part.

SAP 2009 uses a default value based upon:

- type
- age; and
- fuel used

This in turn is used within SAP 2009 to determine the SAP winter (heating and hot water load) and summer (water load only) seasonal efficiency.

- this SAP rating (scale) is between 1-100, where 1 denotes a very poor standard of energy efficiency and 100 is exceptionally high

Seasonal Efficiency of Domestic Boilers in the United Kingdom (SEDBUK)

SEDBUK ratings classify appliances for their average efficiency - in terms of fuel consumed in return for heat produced - and are designed to offer a true comparison as to relative efficiency.

Since its introduction in 1999, SEDBUK has been through a number of changes – recently SEDBUK 2005, to the latest version SEDBUK 2009.

Correspondingly, SAP 2005 uses data from SEDBUK 2005 and SAP 2009 use that of SEDBUK 2009.

The SEDBUK rating is calculated considering:

- the boiler type
- ignition
- the UK climate
- fuel usage patterns, as well as
- standard laboratory tests

Under SEDBUK and up until October 2010 gas appliance were rated via efficiency bands – Band A (90% + efficiency) down to the lowest Band G (below 70% efficiency) – this banding system has been withdrawn so as not to cause confusion with the proposed Energy using Products Directive (EuP), which will introduce appliance labeling in the same vein as that currently used for 'white goods'.

Appliance efficiency is now stated only as a percentage.

Wet gas central heating appliances (boilers) installed in newly constructed buildings and buildings that have been refurbished should have a SEDBUK gross efficiency rating of:

- 90% efficiency based on the older SEDBUK 2005, or
- 88% efficiency based on SEDBUK 2009

Note 1: If the SEDBUK rating is not stated as being 2009, but simply SEDBUK, then it will be assumed to be SEDBUK 2005.

Note 2: Range cooker boilers are allowed a slightly lower energy efficiency rating of 75% – but these appliance types are not discussed in this manual.

However, if an assessment of the proposed installation in an existing dwelling – using the 'Guide to condensing boiler installation assessment procedure for dwellings' – shows that a condensing boiler cannot be fitted, a lower efficiency boiler can be installed.

The minimum efficiency allowed is:

- 78% for Natural gas appliances; and
- 80% for Liquefied Petroleum Gas (LPG) appliances

See the Domestic Building Services Compliance Guide for further guidance.

- the necessary controls must support the appliance in achieving the full SEDBUK performance
- the dwelling needs temperature controls - and provisions are also made for the 'heat exchange' efficiency and adequate insulation for domestic hot water storage vessels
- pipework in unheated areas, including those within the building fabric (floor/wall system, where practicable) and those connected to hot water storage vessels (all pipes from the point of connection to at least 1m) needs to be adequately insulated
- when you refurbish, you are recommended to upgrade gravity hot water systems to a fully pumped design - with adequate temperature control added

Note: Although this is the preferred option (and it would be good practice to upgrade an existing 'gravity fed' circuit to 'fully pumped) you can retain it if replacing proves impractical.

But you would need to upgrade the controls:

- to include a cylinder thermostat and zone valve to control the water temperature of the hot water circuit; and
- to provide an interlock with the boiler to prevent burner cycling

Notification of building work under ADL

Under the regulations, building work must be pre-notified to an appropriate Building Control Body – but if the business is a member of a competent persons' scheme the work is notified to the provider after completion. They in turn notify the relevant building control department.

If you are installing a gas appliance and are competent to do so under the amended Building Regulations, you do not need to notify Building Control before you carry out the work.

Approved Document P (ADP) – points to note

- All electrical work associated with the installation of an appliance and its controls needs to comply with manufacturer's instructions and BS 7671: 2008 (Including Amendment No.1): 2011 Requirements for electrical installations. IET Wiring Regulations Seventeenth Edition
- electrical work needs to be undertaken by suitably competent persons and, where to be self-certified as complying with the Building Regulations, by a member of a competent person scheme
- suitable documentation – electrical installation certificates or minor works certificates – needs to be completed upon completion of works and presented to the user of the installation

Conversion to burn another gas

Carry out conversion to another gas, if necessary, strictly in accordance with the appliance manufacturer's instructions and the kit of parts supplied by them.

Open-flued boilers – 2

Introduction

The popularity of the open-flued boiler configuration meant a wide choice of appliance type – from floor standing models (either stand alone or as part of a combined gas fire and back boiler arrangement – see **Part 6 – Combined gas fire/back boiler**) or wall hung.

They are also included in some cooking ranges – not discussed within this manual – where they typically serve the hot water demand for the user(s).

In the past open-flued gas boilers may have also been regarded as a more flexible configuration when choosing a new gas boiler as they could be installed in most rooms (see **Restricted locations – for safety reasons** in this Part) within a dwelling and did not rely on having access to an outside wall as was the case for room-sealed boilers (see **Part 3 – Room-sealed boilers**); this, as we know is no longer the case.

They do however, have some drawbacks – having the combustion circuit open to the room in which the appliance is installed is perhaps the biggest concern as any problem with the combustion process – blocked/restricted combustion circuit (heat exchangers, baffles, etc) or problems associated with the flue (see **Changes to industry standards – note new definitions** in this Part) – will lead to products of combustion (POC) being discharged back in to the room concerned.

This starts a chain reaction whereby vitiated air (contaminated air) is drawn back in to the appliance for the combustion process, further promoting poor combustion and thereby generating ever-higher levels of Carbon Monoxide (CO).

Note: Correct installation combined with regular maintenance will greatly reduce any safety related issues however, even then open-flued gas appliances can still suffer from atmospheric conditions and/or a phenomenon of thermal inversion (see the current Essential Gas Safety – Domestic – Parts 13 and 14 for further guidance).

Additionally, correctly sized, configured and located ventilation is of critical importance to the continued safety of an open-flued appliance (see **Part 8 – General installation details – Wet central heating – Ventilation – essential for safe operating**).

As part of your survey when considering a new boiler installation ensure that:

- the boiler output is capable of satisfying the heating and/or hot water demands
- existing control systems (in case of boiler replacement) meet the needs of the new boiler and those of ADL1A or ADL1B, as appropriate (see **Part 1 – Central heating wet – Building regulation requirements**)
- the primary circuit is correctly cleansed prior to the new boiler being installed – BS EN 14336: 2004 'Heating systems in buildings. Installation and commissioning of water based heating systems' provides additional guidance. See also **Part 8 – General installation details – Wet central heating**

Where the boiler is to be used on a sealed system, the boiler you select must be specifically designed for this purpose and must incorporate the manufacturer's protection devices for use on sealed systems (see **Part 8 – General installation details – Wet central heating – Sealed system**).

Carbon monoxide alarms

To provide additional peace of mind and combined with correct installation and on going periodic maintenance of gas appliances, the installation of an approved CO alarm – one complying with BS EN 50291-1: 2010 'Electrical apparatus for the detection of carbon monoxide in domestic premises. Test methods and performance requirements' – is recommended.

- ensure the alarm is located as per manufacturer's instructions
- is within its service life (typically 5 – 7 years)
- displays appropriate product conformity i.e. BS EN 50291 and CE marking
- is tested frequently by both the gas user and you when attending the property, by depressing the test button (similar to that required for smoke alarms)

A single CO alarm should be used for one appliance/area. Where multiple open-flued appliances are used or where multiple rooms such as bedrooms are required to be monitored, separate CO alarms should be installed at each appliance/in each area – note that some areas such as kitchens will be unsuitable for standard domestic CO alarms due to cooking activities in that area (consult with the CO alarm manufacturer should this area be under consideration).

For further guidance on CO alarms see the current Essential Gas Safety – Domestic – Part 3.

Changes to industry standards – note new definitions

BS EN 1443: 2003 'Chimneys. General requirements' provide some new definitions (found in BS 5440-1: 2008 'Flueing and ventilation for gas appliances of rated input not exceeding 70 kW net (1st, 2nd and 3rd family gases). Specification for installation of gas appliances to chimneys and for maintenance of chimneys'), which alter common terms we have been used to within the gas industry, when we talk about flues and chimneys:

Chimney: This is a structure consisting of a wall or walls enclosing the flue or flues.

Chimney component: This is any part of a chimney.

Flue liner: This is the wall of a chimney consisting of components, the surface of which is in contact with the POC. This is not just a flexible flue liner, but also any suitable material to convey these products.

Flue: This is the passage or space for conveying POC to the outside atmosphere.

Therefore, what we have previously referred to as a 'flue', particularly for room-sealed configurations, is now classed as a chimney unless we are discussing the passage of POC.

Restricted locations – for safety reasons

Basements and cellars

- do not install a boiler fitted with an automatic means of ignition for use with LPG in a room or internal space below ground level e.g. a basement or cellar
- you can install such boilers into rooms which are basements with respect to one side of the building, but open to ground level on the opposite side

Bath or shower rooms

- do not install open-flued gas boilers in a room or internal space containing a bath or shower

  This includes any cupboard/compartment or space (e.g. cubicle), which has an air path or connecting door opening into the bath or shower room.
- only install room-sealed appliances – only consider this location if there is no alternative location and only if the manufacturer's installation instructions allow this (see **Part 8 – General installation details – Wet central heating – Gas appliances in bathrooms – electrical zoning requirements**)
- you may service or repair an existing open-flued gas boiler installed before 24th November 1984, in such a location, provided it is **safe to use**, but;
  - you must class it as a Not to Current Standards (NCS) installation, in accordance with the current Gas Industry Unsafe Situations Procedure (see also the current Essential Gas Safety – Domestic – Parts 8 and 10)

Bedroom/bedsitting rooms

- gas boilers of greater than 14kW (gross) heat input (12.7kW net) installed in a room used or intended to be used as sleeping accommodation must be room-sealed
- this includes the installation of a gas appliance in any cupboard/compartment or space (e.g. cubicle), which has an air path or connecting door opening into the bedroom/bedsitting room
- gas boilers of 14kW (gross) heat input (12.7kW net) or less may be room-sealed, or;
- if open-flued, must incorporate a safety control designed to shut down the appliance before there is a dangerous quantity of products of combustion (POC) in the room concerned. This device should be in the form of an atmosphere sensing device (see **Part 13 – Definitions**).

Note: These requirements apply to new installations including used or second-hand gas appliances installed after the 1st of January 1996. Existing appliances in these locations, provided they are safe to use, may be serviced or repaired, but should be classed as a NCS installation in accordance with the current Gas Industry Unsafe Situations Procedure.

The use of a fixed CO detector/alarm, conforming with BS EN 50291-1 and bearing a CE/Kite mark is also recommended
(see **Carbon monoxide alarms** in this Part).

Covered passageways – make sure no POC

- do not site chimney terminals within a covered passageway between properties e.g. terraced properties
- existing chimney systems terminated in these positions will produce POC, which could contain CO and could accumulate entering habitable areas above the passageway

Note: When a gas operative encounters an existing gas appliance chimney terminal within a covered passageway, the gas operative will need to carry out a risk assessment of the installation in accordance with the procedure outlined in Gas Safe Registers Technical Bulletin 007 'Room-sealed (natural draught) and open-flued (fanned draught) appliance flues terminating within covered passageways (ginnells)' – visit https://engineers.gassaferegister.co.uk

Private garages – points to note

The Gas Safety (Installation and Use) Regulations 1984 banned the installation of open-flued appliances in private garages to reduce the risk of explosion/fire from hazardous substances such as petroleum vapour. This ban was relaxed on 31st October 1994.

The Building Regulations in Scotland also placed restrictions on this type of installation, but this ban has also been relaxed and a gas appliance in a garage no longer has to be room-sealed.

- if you install an open-flued boiler in a private garage, follow the appropriate regulations
- the majority of garages will be unheated spaces, therefore adequate frost protection will need to be installed
- also note that some manufacturers do not allow this type of installation
- where you find an existing open-flued gas boiler installed in a private garage, advise the customer to check their insurance policy (which may be affected by this type of installation)

Protected shafts/stairway – prevent the spread of fire

Protected shafts are stairs or other shafts passing directly from one compartment floor to another, enclosed so as to prevent the spread of fire or smoke.

A protected shaft can often be found in flats over two storeys high, with individual accommodation on each floor, where the means of access and exit is via the protected stairway.

- no gas appliances are permitted in a protected shaft/stairway

Installation

To avoid repetition, general information will be found in **Part 8 – General installation details – Wet central heating**, including details on boiler locations, gas supply and ventilation.

General

- open-flued boilers covered by this Part can be floor-standing or wall-mounted
- you must connect them to the gas supply by a permanently fixed and correctly sized rigid pipe
- the final connection to the boiler must incorporate an isolating tap and a means of disconnection to facilitate removal for servicing/maintenance etc.
- you must connect the boiler to a permanent chimney system and design the chimney in accordance with the manufacturer's instructions (see also the current Essential Gas Safety – Domestic – Part 13 for further guidance)

Air supply – what you must check

- all open-flued gas boilers need air:
  - for combustion
  - for cooling, when installed in an enclosure/compartment; and
  - to assist the safe operation of the flue
- size the ventilation opening and test the chimney system in accordance with the manufacturer's instructions (see also the current British Standard for ventilation requirements – BS 5440-2: 2009 'Flueing and ventilation for gas appliances of rated input not exceeding 70kW net (1st, 2nd and 3rd family gases) – Part 2: Specification for the installation and maintenance of ventilation provision for gas appliances' and the current Essential Gas Safety – Domestic – Part 4 for further guidance)

Open-flued chimneys

- before you begin installing an open-flued central heating boiler to an existing chimney system, verify the correct operation of the flue (see the current Essential Gas Safety – Domestic – Part 14 for further guidance)
- also refer to the manufacturer's installation instructions to check that the boiler is suitable for the chimney type you are connecting it to, including its method and location with regards to termination of the chimney system

Room-sealed boilers – 3

Introduction

Modern central heating boiler technology allows increased flexibility in room-sealed appliances, so you are always recommended to consider installing these instead of an open-flued type.

If there is no outside wall to site a room-sealed terminal, consider installing a fanned draught room-sealed boiler with a vertical chimney system option (see Note).

Note: European Standards have brought about a change to terminology used in relation to 'Flues' – see Part 2 – Open-flued boilers – Changes to industry standards – note new definitions.

Advancement in boiler technology has greatly increased the flueing options available to the gas installer, to such an extent that it is highly unlikely that a dwellings nature of construction will preclude the installation of a suitable room-sealed system.

Efficiency drives mean that most modern boilers are of the fanned draught type, this allows for longer chimney runs and in the case of condensing models (whereby the flue gas temperature is low), smaller diameter flues and air intakes, as well as using lighter materials in their construction such as PVC.

Chimney configurations can be of the concentric (flue pipe within the air intake pipe) or twin pipe design and can run over a considerable distance within a structure to the point of termination in the outside air.

As part of your survey when considering a new boiler installation ensure that:

- the boiler output is capable of satisfying the heating and/or hot water demands
- existing control systems (in case of boiler replacement) meet the needs of the new boiler and those of ADL1A or ADL1B, as appropriate (see **Part 1 – Central heating wet – Building regulation requirements**)
- the primary circuit is correctly cleansed prior to the new boiler being installed – BS EN 14336 provides additional guidance. See also **Part 8 – General installation details – Wet central heating**

Where the boiler is to be used on a sealed system, the boiler you select must be specifically designed for this purpose and must incorporate the manufacturer's protection devices for use on sealed systems (see **Part 8 – General installation details – Wet central heating – Sealed system**).

Restricted locations – for safety reasons

Basements and cellars

- you must not install a boiler fitted with an automatic means of ignition for use with LPG in a room or internal space below ground level e.g. a basement or cellar
- you may install such boilers into rooms which are basements with respect to one side of the building, but open to ground level on the opposite side

Covered passageways – make sure no POC

- do not site chimney terminals within a covered passageway between properties e.g. terraced properties
- existing chimney systems terminated in these positions will produce POC, which could contain CO and could accumulate entering habitable areas above the passageway

Note: When a gas operative encounters an existing gas appliance chimney terminal within a covered passageway, the gas operative will need to carry out a risk assessment of the installation in accordance with the procedure outlined in Gas Safe Registers Technical Bulletin 007 'Room-sealed (natural draught) and open-flued (fanned draught) appliance flues terminating within covered passageways (ginnells)' – visit https://engineers.gassaferegister.co.uk

Protected shafts/stairway – prevent the spread of fire

Protected shafts are stairs or other shafts passing directly from one compartment floor to another, enclosed to prevent the spread of fire or smoke.

A protected shaft is often found in flats over two storeys high, with individual accommodation on each floor, where the means of access and exit is via the protected stairway.

- no gas appliances are permitted in a protected shaft/stairway

Installation

To avoid repetition, general information will be found in **Part 8 – General installation details – Wet central heating**, including details on boiler location and gas supply.

General

- room-sealed boilers covered by this Part can be either floor standing or wall hung
- you must connect all room-sealed boilers to the gas supply by a permanently fixed and correctly sized (see Note) rigid pipe

Note: Correctly sized gas pipework is crucial for correct/efficient operation of modern gas boilers – even more so for combination boilers as these will be adversely affected, in terms of poor performance if the pipework is undersized (see Part 4 – Combination boilers).

- the final connection to the boiler must incorporate a gas isolating valve and means of disconnection - to facilitate removal for servicing/maintenance etc.

Ventilation

Additional purpose provided ventilation is not required for combustion purpose with room-sealed appliances. However, ventilation for cooling purposes maybe required for those appliances installed in compartments/enclosures.

Refer to the manufacturer's installation instructions for further guidance – note that some appliances will not require cooling ventilation to be provided due to the nature of their design.

Room-sealed chimneys

- do not locate the terminal of a room-sealed or a fanned draught room-sealed boiler over neighbouring property
- do not locate a terminal so that POC are discharged across a neighbouring boundary
- ensure the terminal position is in accordance with the manufacturer's instructions and BS 5440-1, in terms of clearances from building features and re-entry points in dwellings (see also the current Essential Gas Safety – Domestic – Part 13 for further guidance)
- chimney's routed through voids (floor, ceiling, etc) have additional safety considerations/installation requirements prescribed by ADL – see **Part 8 – General installation details – Wet central heating – Chimneys in voids** for further guidance

Figures

Introduction

Unlike a 'traditional' central heating boiler, a combination boiler is produced as a complete package, incorporating all the components required to operate and control a full central heating system - as well as the ability to provide instantaneous hot water to a number of draw-off taps.

The compact nature of combination boilers is very attractive to both gas users and gas installers alike, as they allow a greater degree of flexibility when deciding a suitable location and, for the installer, generally quicker/less disruptive installation.

Having all the components contained within the boiler chassis (with the exception of the heating thermostats), as well as 'on demand' hot water supply means that a 'traditional' airing cupboard or similar enclosure/compartment is not required – particularly attractive to developers of multi-occupancy dwellings (flats) given the premium placed on space – nor is there the need for CWSC and associated pipework in the roof space.

Combination boilers can however, supply a bulk store of hot water should the design call for this, which will mean additional controls including a hot water storage vessel will be required as well as the space needed to house them – but this would negate a major selling point of installing a combination boiler to start with.

It is important to remember with combination boilers that adequate water pressure and flow are essential:

- before you install, check availability of a water supply of adequate pressure and flow

Combination boilers use cold water directly from the mains to supply the hot water taps. The rising water main must also be able to provide other cold water outlets simultaneously.

- if you fail to take this into account, the appliance may perform unsatisfactorily, as it requires an adequate water pressure and flow rate to activate various controls; and
- lead to customer complaints of poor performance

The manufacturer will provide details of the minimum pressure required, but a typical figure is between 1 and 1.35bar.

Note: You must ensure the gas supply is adequate for the purpose (see 'Installation – General points to note and follow' in this Part).

As previously mentioned, combination boilers are flexible and such can be installed in most locations within a dwelling however, there are still restrictions on some areas depending on the appliance configuration – see **Part 2 – Open-flued boilers** and **Part 3 – Room-sealed boilers – Restricted locations – for safety reasons** – for further guidance on 'restricted locations'.

Installation

To avoid repetition, general information will be found in **Part 8 – General installation details – Wet central heating**, including details on boiler location, gas supply and ventilation.

General points to note and follow

When you install a combination boiler (whether open-flued or room-sealed, wall mounted or floor standing), follow this manual's general guidelines for open-flued boilers and room-sealed boilers.

Most combination boilers have a high gas input rate (typically 32kW (gross) or greater) so as to provide a reasonable water flow rate with a 35°C temperature rise. To do this, they must have an adequate gas supply.

However, most 32kW input (gross) combination boilers are fitted with a multifunctional gas control valve having ½ inch BSP connections and a ½ inch BSP isolating valve.

- this suggests that 15mm pipework is sufficient to meet this demand - although the length of 15mm pipework is generally used at the point of connection and is typically limited to within 1m of the boiler
- you may have to increase the remaining pipework back to the gas meter to 22mm, 28mm - or in some cases 35mm depending on the load and length of pipe run (see the current Essential Gas Safety – Domestic – Part 5 for further guidance)
- where you use a tube cutter to cut the pipe, you must suitably de-burr the pipe ends. Removing the burr(s) reduces restriction and turbulence within the pipe and maintains gas flow
- failure to provide an adequate gas supply is likely to result in both reduced performance from the boiler - and failure to satisfy user expectations
- you must connect the boiler to the gas supply by a permanently fixed, correctly supported and sized rigid pipe
- ensure the final connection to the boiler incorporates an isolating valve and means of disconnection to facilitate removal for servicing/maintenance etc.
- also ensure that if the combination boiler replaces an existing 'standard design' central heating boiler or water heater, the existing gas supply is adequate to provide for the combination boiler - based on the highest gas input rating e.g. the hot water usage

  If you use an existing 'plugged point' in the gas supply designed for another appliance (for example, a gas cooker) then it's doubtful it will satisfy the demands of a combination boiler.

The basic components of a combination boiler are shown in Figure 4.1.

Note: Other appliance manufacturers' utilise differing design philosophies such as separate heating and hot water circuits passing through the same heat exchanger – and because of this will have fewer controls than those shown in Figure 4.1.

Sealed system

Because of the pressurised nature of sealed systems, as opposed to open-vented systems, take extra precautions to ensure the installation complies with the current Water Supply (Water Fittings) Regulations
(see **Part 8 – General installation details – Wet central heating – Sealed system**).

Figure 4.1 Basic components of a combination boiler

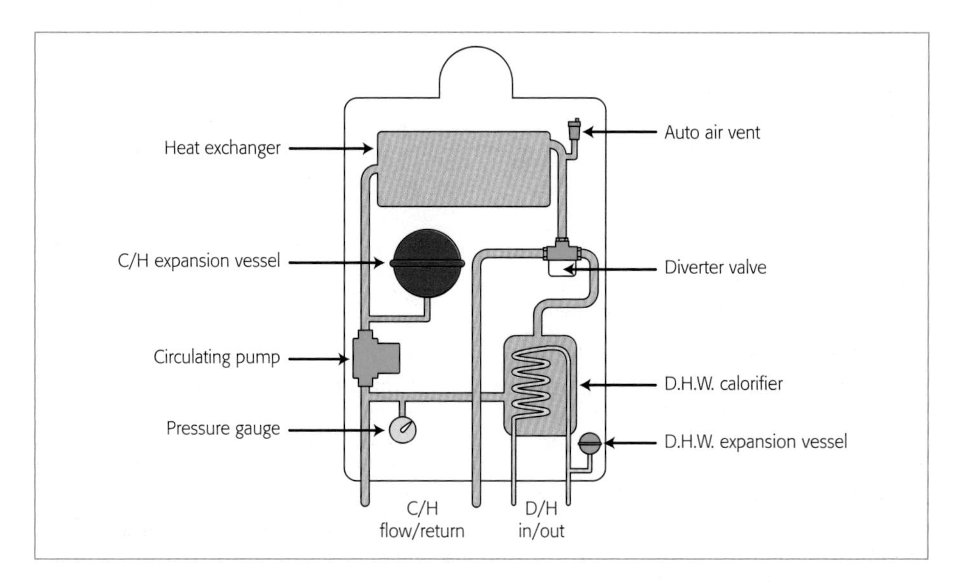

Domestic hot water connection

- you must connect the domestic hot water side of the combination boiler to the cold water main
- where the cold water inlet supply has a non-return valve (NRV) fitted e.g. a water meter or stopcock with a loose jumper valve, then fit the water supply with a suitable expansion vessel
- fit the vessel between the boiler and valve to accept the expansion of domestic hot water from the boiler. Some boiler manufacturers include this expansion vessel as part of the boiler components. It's generally the size of a tennis ball (see Figure 4.1)

Water supply areas – differing characteristics

Unlike traditional central heating boilers that heat water, which is then stored in a hot water storage vessel, combination boilers heat the water 'on-demand' – combination boilers can be configured to supply a storage vessel, but this isn't a common practice (other boiler types are more suited to this set-up).

- in 'temporary hard' water areas, take precautions to ensure lime scale build-up doesn't reduce the efficiency of the domestic hot water calorifier

  ADL recognises the issues related to hard water – 200ppm and above – and where required by the appliance manufacturer, calls for suitable treatment to limit the formation of lime scale
  (see **Part 1 – Central heating wet – Building regulation requirements – Approved Document G – points to note – Hard water areas** for further guidance)

- to reduce this risk, most combination boilers now incorporate a plate heat exchanger (claimed to reduce scale build-up)
- in all cases, carefully follow the boiler manufacturer's installation instructions

Note: Water is said to be 'hard' when ordinary soap doesn't produce an immediate lather. The water contains calcium sulphate and remains permanently hard; you cannot remove the hardness by boiling.

'Temporary hardness' occurs in water that contains calcium bicarbonate, which you can remove by heating the water. This changes the soluble calcium bicarbonate into the insoluble calcium carbonate. As a consequence, the calcium carbonate is deposited in the heat exchanger in the form of 'scale'.

In hard water areas, further guidance from the manufacturer may be required to combat associated scale issues.

Condensing boilers – 5

Figures

Introduction

Gas installers are very familiar with modern condensing gas boilers – given their proliferation over recent years – offering, as they do greater efficiency than any other type of gas boiler and as an added benefit of their mode of operation, have greater flexibility when it comes to installation.

As most of the 'useful' (sensible and latent) heat is recovered from the POC – see **System design – standard boiler and condensing boiler** in this Part – the POC are at a very low temperature allowing appliance manufacturers' to design chimney systems which are:

- lighter than traditional metallic systems, utilising materials such as PVC
- reduced diameter
- varied design – use of twin pipe design (separate air inlet and flue pipe) is very common along with more 'traditional' concentric chimney designs (pipe within a pipe); and
- consequently, can run over greater distances to the point of termination

Note: The last bullet point is of particular interest, as chimney systems can be routed through various building elements, which in the past may not have been an option. These elements include floor, wall and ceiling voids, which although acceptable, do have additional requirements placed upon them by ADL – see 'Part 8 – General installation details – wet central heating – Chimney's in voids' – for further guidance.

Before installing a condensing gas boiler, check during your survey that the boiler output can satisfy the heating and/or hot water demands.

Where the boiler is to be used on a sealed system, the boiler you select must be specifically designed for this purpose and must incorporate the manufacturer's protection devices for use on sealed systems (see **Part 8 – General installation details – Wet central heating – Sealed system**).

Installation

To avoid repetition, you will find general information in **Part 8 – General installation details – Wet central heating**, including details on boiler location and gas supply.

General

- when you install a condensing boiler (whether wall-mounted or floor-standing) follow the general guidelines for room-sealed boilers, outlined in **Part 3 – Room-sealed boilers**
- connect all gas boilers to the gas supply by a permanently fixed, correctly supported and sized rigid pipe
- the final connection to the boiler must incorporate an isolating valve and means of disconnection to facilitate removal for servicing/maintenance etc.

Figure 5.1 Condensing boiler

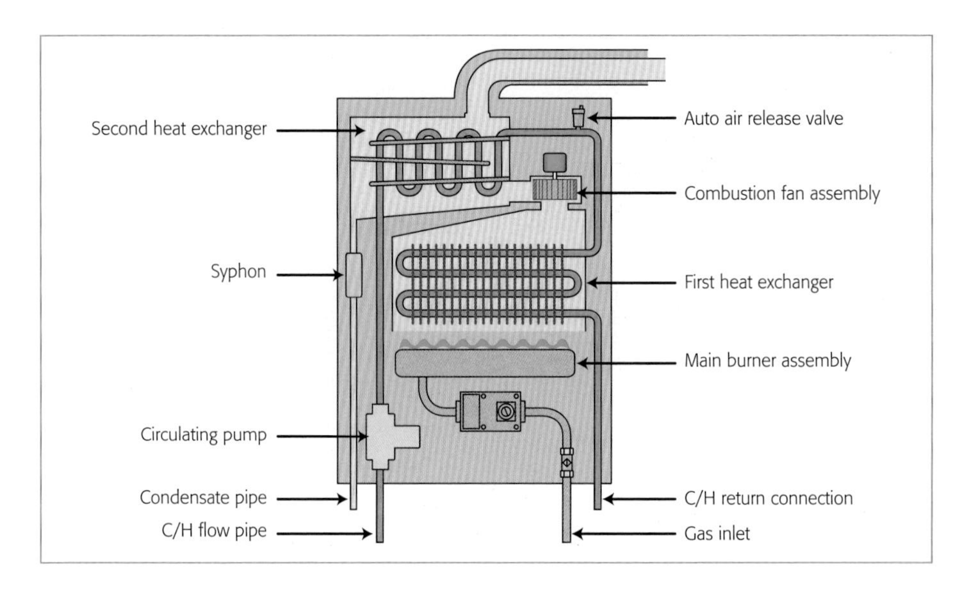

System design – standard boiler and condensing boiler

In a standard boiler, heat is transferred from the hot POC to the heat exchanger containing the system water.

To prevent condensation forming, the boiler heat exchanger is designed so that the POC do not fall below 55°C. This stops condensation - which could cause corrosion and damage the boiler. Also, the reduced heat available would adversely affect flue performance.

The principal difference between a 'standard' boiler and a condensing boiler is that the heat contained in the water vapour of the POC is recovered.

A two-phase, single extended surface area heat exchanger or a double pass heat exchanger achieves this. Figure 5.1 shows a typical arrangement.

A condensing boiler recovers two types of heat:

Sensible heat

This means the heat extracted from the POC, as they pass through the first phase or first pass heat exchanger, where the POC temperature is reduced.

Latent heat

POC contain water in the form of a vapour because of the heat of the POC. Heat is recovered from this water vapour by additionally passing the POC through the second part of the heat exchanger.

- to get maximum advantage from the increased efficiency, condensing boilers need to be kept in condensing mode by designing the heating system to give a flow/return temperature difference of 21°C (as opposed to the 11°C for a standard central heating system)

Figure 5.2 Baxi GasSaver

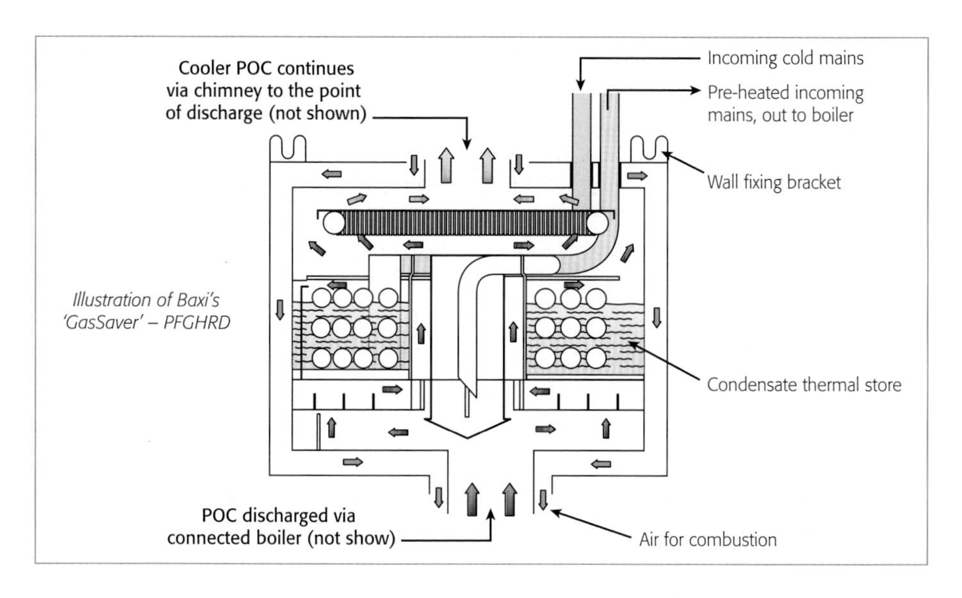

Further advancement in condensing technology has seen the introduction of Passive Flue Gas Heat Recovery Devices (PFGHRDs), which when combined with a condensing boiler offer even greater efficiency gains – See **PFGHRD – enhancing existing technology** – in this Part.

PFGHRD – enhancing existing technology

Being conceptualised and refined over a number of years, Zenex Technologies has brought to the gas market 'passive flue gas heat recovery devices' – PFGHRDs, which in essence scrub the remaining heat within the POC that in turn is used to pre-heat the mains water supply to the boiler.

By pre-heating the incoming water supply, the boiler heats the required water demand quicker and therefore, uses less energy and this increases the efficiency of the appliance even further.

PFGHRDs are a self-contained unit having no moving parts, which is connected above the boiler on to the appliance flue spigot. A water inlet (connected to the cold mains) and outlet, which connects to the boiler cold feed inlet, is incorporated into the design (See Figure 5.2).

As POC is discharged by the boiler, it passes through a third heat exchanger contained within the unit, scrubbing the remaining heat from the POC.

The incoming mains water is first passed through this third heat exchanger, pre-heating the water within. As the POC is stripped of its remaining heat, it condenses within the unit forming a 'pool' of condensate, which still has a little heat remaining in it.

A water way from the heat exchanger is passed through this pool (condensate thermal store) extracting the last ounce of available heat.

This water, being pre-heated then feeds the boiler to be heated fully to its operating temperature. The POC then exits the unit to be discharged by the chimney system in the normal fashion.

With so much moisture being extracted from the POC, the problem of 'pluming' – an issue associated with condensing boiler technology in particular – is all but eliminated. Good news for customers and gas installers alike.

The technology has been embraced by various boiler manufacturers', who in turn have developed their own variants of the design, offering tighter product integration.

PFGHRDs are recognised by SAP 2009 – see **Part 1 – Central heating wet – Building regulation requirements – Approved Document L (ADL1A and ADL1B) – points to note – Standard Assessment Procedure (SAP)** – and therefore contribute to the dwellings energy efficiency rating.

Condensate discharge

A condensing boiler produces a significant quantity of condensate (a 30kW (gross) boiler in constant operation produces about 4 litres of condensate per hour), which needs to be removed from the boiler and discharged in to the foul system of the dwelling.

BS 6798: 2009 'Specification for installation and maintenance of gas-fired boilers of rated input not exceeding 70kW net' provides guidance on suitable methods of discharging condensate, which are:

- via an internal or external soil and vent stack
- via an internal waste pipe
- via an external drain or gully
- via a rainwater hopper that is part of a combined system – connected to a sewer which carries foul and rainwater; or
- via a purpose designed soakaway

The majority of the discharge options rely on gravity discharge, so it's important that the pipework is installed with the correct fall – minimum 2.5° to the horizontal (approximately 45mm in every metre) – so as to avoid condensate being trapped.

Where gravity cannot be relied upon, the use of a condensate pump needs to be employed.

The various methods of discharging condensate will, in large part be dictated by the dwelling/installation in question. However, there is a preferred order in which the discharge should be chosen – that order is:

1). Internal discharge under gravity to the waste pipe, soil and vent stack.

2). Internal discharge using a condensate pump to the waste pipe, soil and vent stack.

3). Where 1 and 2 cannot be employed, discharge externally to a waste pipe, soil and vent stack, drain/gully, or purpose designed soakaway.

Note: Option 3 requires additional installation provisions, which will be discussed in this Part.

Internal discharge – under gravity

The preferred choice for all installations as the discharge is protected from freezing conditions – a real issue in recent years given the severe weather the UK has experienced – and connects to the internal soil and vent stack (see Figure 5.3) or waste pipe from sink, basing, bath or shower (see Figures 5.4 and 5.5).

Connection via soil and vent stack:

- unrestricted length of condensate discharge
- minimum pipe diameter is 22mm plastic
- continuous fall of at least 2.5° to the point of connection to the waste system
- supported (pipe clips) – 0.5m for horizontal runs and 1m for vertical runs

Figure 5.3 Condensate pipe connected to internal soil and vent stack

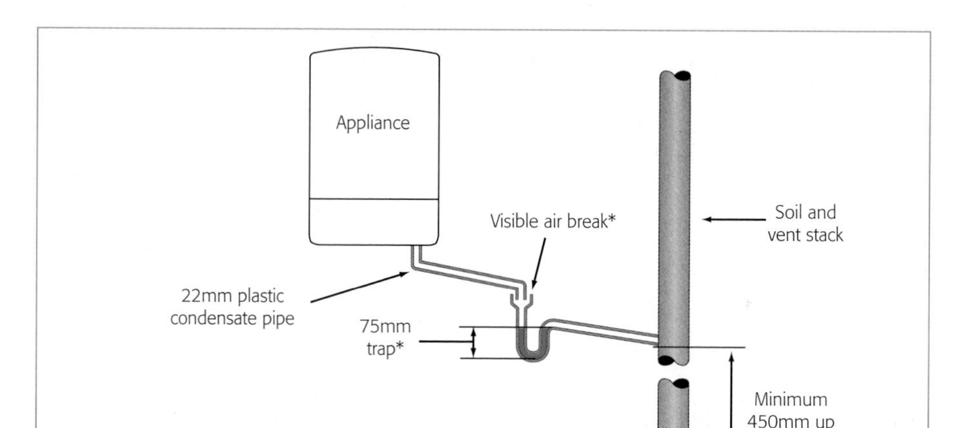

- installation of a 75mm trap with air break (see Note)
- connected a minimum 450mm above the bend at the foot of the stack (up to 3 storeys – distance increases on taller structures), or if unseen, measured from where it can be seen

Note: Where appliances incorporate a trap with a minimum condensate seal of 75mm, a separate trap and air break is not required – consult manufacturer's instructions for further guidance.

Connection via waste pipe:

- unrestricted length of condensate discharge
- minimum pipe diameter 22mm plastic
- continuous fall of at least 2.5° to the point of connection to the waste system
- supported (pipe clips) – 0.5m for horizontal runs and 1m for vertical runs
- condensate connection downstream of sink, basin, bath or shower trap – 75mm trap and air break required, unless the appliance has an internal condensate seal of 75mm or more
- condensate connection upstream of sink, basin, bath or shower trap – an air break is required. Where the sink, basin, bath or shower has an integral overflow; this will serve as the visible air break
- minimum 100mm between the top of the sink, basin, bath or shower to the condensate trap of the appliance, or the additional condensate trap outside the appliance, as appropriate

Figure 5.4 Condensate pipe connected to internal waste pipe – downstream

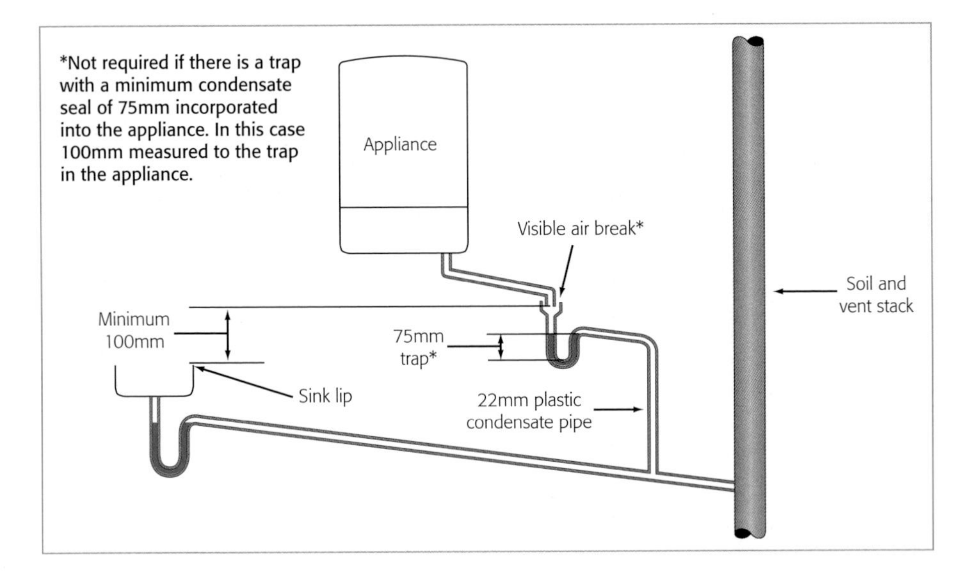

Figure 5.5 Condensate pipe connected to internal waste pipe – upstream

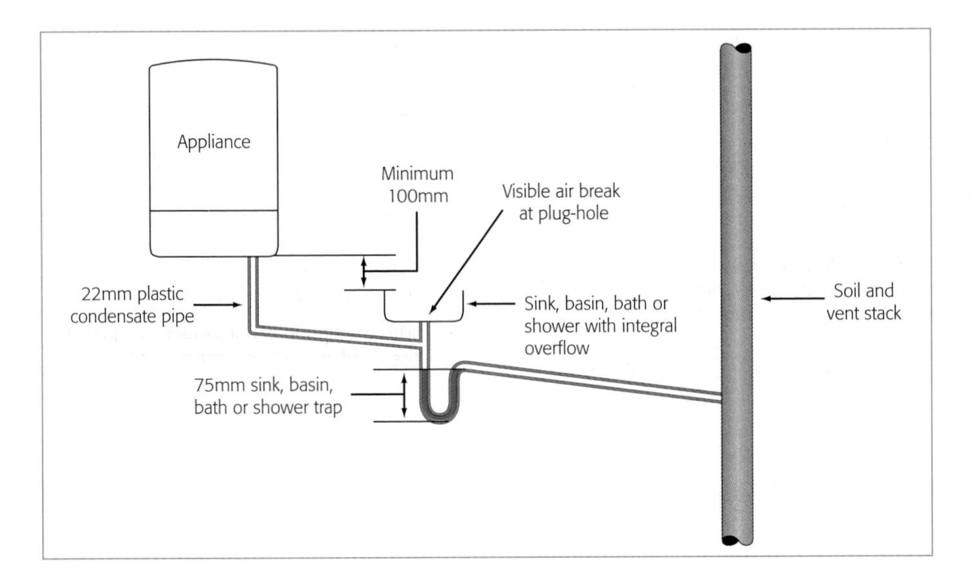

Figure 5.6 Condensate pipe connected and discharged via condensate pump

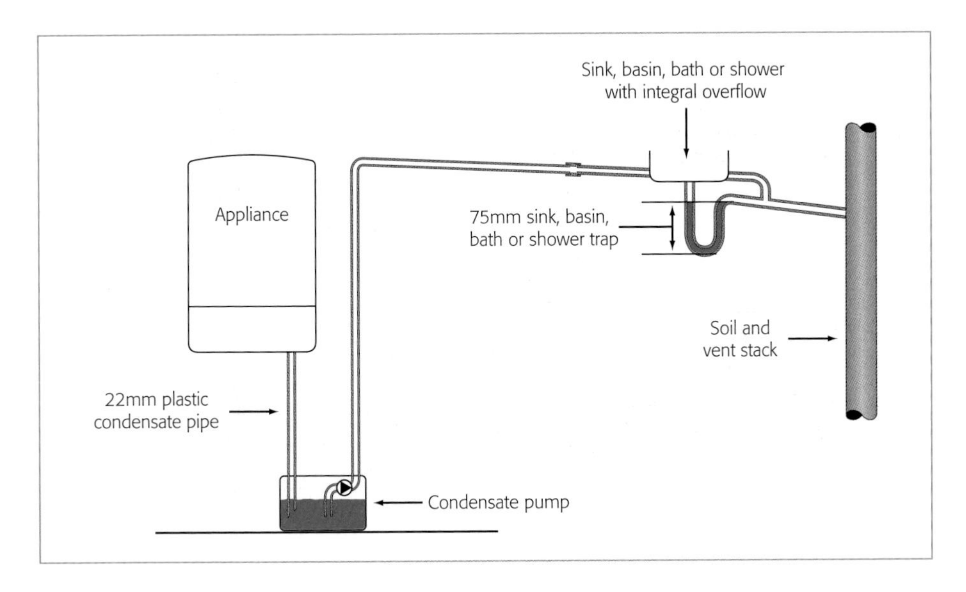

Internal discharge using condensate pump

Where a boiler is to be installed lower than any available internal waste system – basement installation for example – or where long internal runs are unavoidable, a condensate pump meeting the manufacturer's specifications needs to be installed (see Figure 5.6).

The outlet of the condensate pump will be connected to the internal waste system, which meets the requirements already discussed for '**Internal discharge – under gravity**' in this Part.

External discharge

If no other internal option is available/practical, it is permissible to run condensate discharge externally to a point of termination – drain/gully, or proprietary soakaway
(see Figures 5.7 and 5.8, respectively).

However, as the condensate is exposed to weather conditions i.e. sub-zero temperatures, a risk of condensate pipes freezing needs to be addressed:

- boilers having a siphon included (designed to remove the condensate once it has reached a predetermined level) as part of the condensate trap arrangement – 22mm pipe may be used up to 3m, thereafter the pipe needs to be insulated (see **HHIC Freezing condensate working group** for additional guidance in this Part)
- where a siphon is not included as part of the condensate trap arrangement – 32mm pipe is required
- any insulation applied needs to be weather resistant and water-proof

Figure 5.7 Condensate discharging externally in to drain

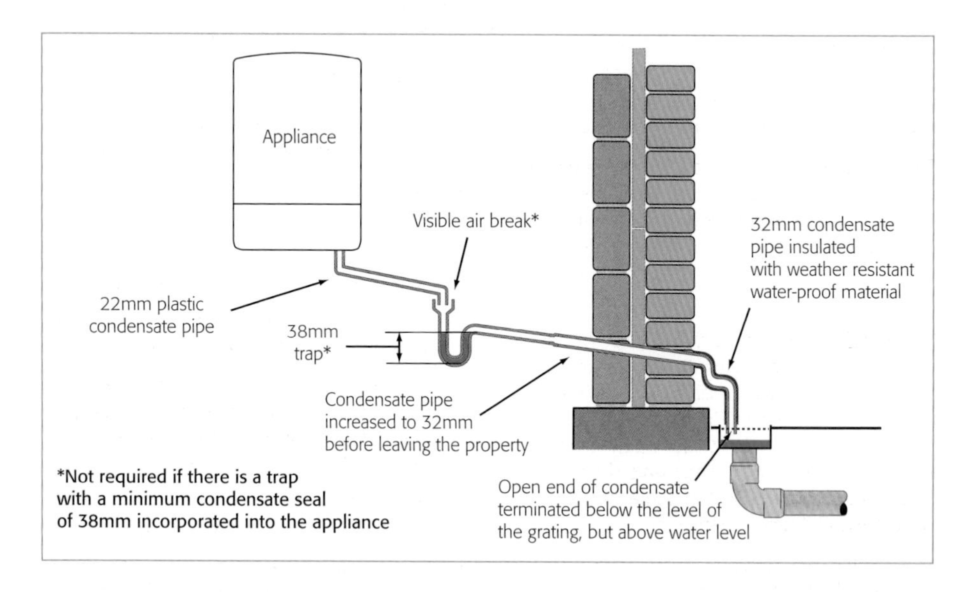

Figure 5.8 Condensate discharging externally in to proprietary soakaway

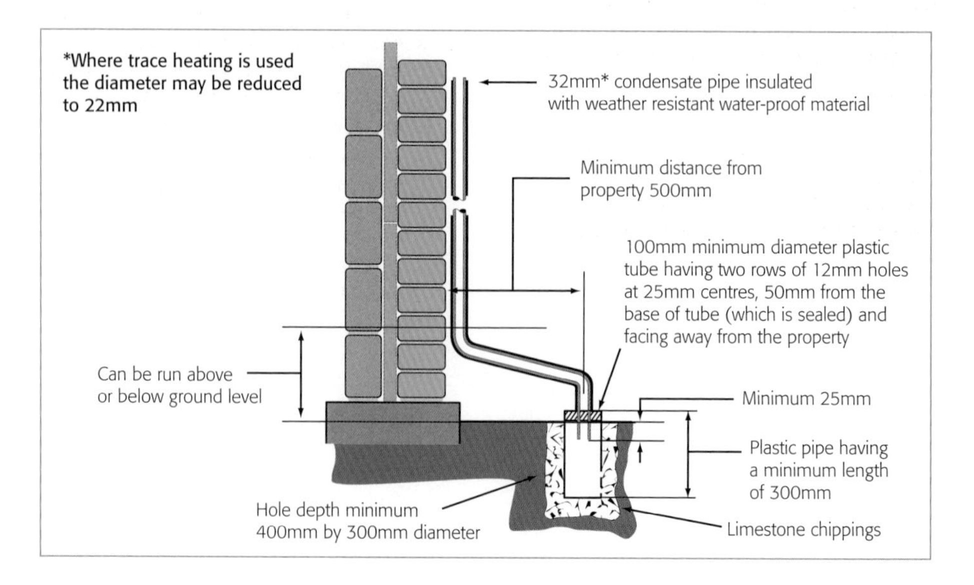

HHIC Freezing condensate working group

Recent weather patterns have seen the UK experience prolonged, sub-zero weather conditions, which have lead to a national problem of freezing condensate pipework.

This resulted in boiler lockout at a time when the heating system is most in need, placing people at risk of exposure (elderly in particular and where no other heat source is available) and at a time when the 'engineering force' is stretched to full capacity.

Some of the issue related to freezing condensate pipework could have been avoided in the first place, had installations complied with the conditions already outlined in this Part. However, problems have been experienced even when the external condensate discharge arrangement complies with manufacturer's instructions and British Standards.

As a result of the problems experienced nationally, the HHIC (Heating & Hotwater Industry Council) set-up a working group to review current practice and information concerning condensate discharge and to identify whether any additional provisions were needed.

For internal installations, the guidance is unchanged to that already discussed – although HHIC have stressed the order of importance when considering condensate discharge (internally under gravity, followed by internally using a condensate pump and finally, externally).

For external installations however, the following additional guidance is provided:

- condensate pipe to run internally as far as possible, upon which the pipe diameter is increased to 32mm before passing through the wall to the exterior of the property
- pipework to take the shortest and least exposed route to the point of discharge
- pipework to fall as steeply as possible away from the boiler – no horizontal runs
- limit the use of fittings within the condensate pipe and deburr pipe before installation
- where terminated over a drain/gully, the pipe should terminate below the grating, but above the water level to minimize 'wind chill' at the open-end
- additional drain covers may also be considered to offer further protection
- condensate pipe to be insulated with suitable weather resistant and water-proof insulation
- on installations where there is likely to be extreme temperatures (exposed sites or as a result of the geographical location), proprietary trace heating incorporating a frost thermostat should be considered in addition to the provisions already highlighted.

  Where trace heating is employed, the requirement to use 32mm pipework does not apply

Combined gas fire/back boilers – 6

Figures

Introduction

A popular choice of heating appliance within the UK and one that has been around for a considerable number of years; the back boiler unit (BBU) and gas fire.

BBU's make use of existing space within a property – a builder's opening – with many of today's existing housing stock being built in the late 19th and early 20th century; new properties can and do have a masonry chimney included in their design, but it's not as prevalent as it once was.

The builders opening is typically constructed of brick/masonry and originally intended for use with solid fuel appliances – having installation features such as 'chair' bricks which channel POC to the throat of the chimney as well as directing radiant heat back into the room, which need to be removed to accommodate a BBU.

As technology developed, BBU's could also be installed in to metallic gas flue boxes, which were later enclosed by suitable building materials to give the appearances of a 'traditional' fireplace.

A gas fire – designed and supplied by the manufacturer as part of the boiler installation – connects to the front of the BBU, again giving the appearance of a traditional fire as well as a secondary heat source in to the room concerned.

The BBU has, until recently – see **Evolution of the design** in this Part – been open-flued requiring either a flexible metallic flue liner (one entire length connecting the flue spigot of the appliance to the terminal of the chimney system) to be inserted in to the chimney or, where the chimney already consists of a clay liner in good working condition, a short flue pipe connecting the appliance to the base of the chimney system, which is then sealed.

Being open-flued there are restrictions on suitable locations, which are discussed in **Part 2 – Open-flued boilers**. For general installation details see **Installation** in this Part.

As with all open-flued gas appliances, adequate ventilation is required to ensure continued safe operation of the appliance – see **Part 8 – General installation details – Wet central heating** and the current Essential Gas Safety – Domestic – Part 4 – Ventilation – for additional guidance.

Before you begin installing a gas fire/BBU, check during your survey that the boiler output can satisfy the heating and/or hot water demands.

Where the boiler is to be used on a sealed system, the boiler you select must be specifically designed for this by the manufacturer – and must incorporate the appliance manufacturer's appropriate protection devices for use on sealed systems (see **Part 8 – General installation details – Wet central heating – Sealed system** – for further guidance).

Evolution of the design

A leading UK manufacturer in BBU technology – Baxi – have further developed the design to make it fully compliant with today's legislative requirements with regards to appliance efficiency.

The 'Baxi Bermuda' (a name that many of us older gas installers will be familiar with) has evolved to be a room–sealed, high efficiency condensing boiler (90+% – previous SEDBUK Band A, see **Part 1 – Central heating wet – Building regulation requirements – Approved Document L (ADL1A and ADL1B) – points to note** in this Part), combined with a 'Valor' electric fire.

Being room-sealed, there is no requirement for additional purposes provided ventilation to be installed and the existing masonry chimney is utilised to contain a concentric chimney system.

The BBU is a self-contained package that includes all the necessary controls, including a condensate pump (see **Part 5 – Condensing boilers – Condensate discharge** – for further guidance) – offering a real choice when looking to replace existing BBUs.

Building Regulations and ADJ

As we've previously looked at in **Part 1 – Central heating wet – Building regulation requirements** – the Building Regulations (England and Wales) 2010 and its supporting documentation in the form of Approved Documents (ADs) sets out the minimum requirements for "securing reasonable standards for health and safety in or around buildings".

In order to secure reasonable standards, building regulations require either:

- to provide the local authority with a building notice in accordance with regulation 13; or
- to deposit full plans with the local authority in accordance with regulation 14

However, 'persons' who carry out the building work are not required to give a building notice or submit full plans where:

- the work as described by Schedule 3 of the regulations is undertaken; and
- by a person, or employed by a person, who is a member of a class of person approved in accordance with regulation 3 of the Gas Safety (Installation and Use) Regulations 1998 i.e. Gas Safe Registered; and
- in the case of 'controlled service and fitting', a person who is a member of a competent person self-certification scheme listed in schedule 3

A controlled service or fitting means "a service or fitting in relation to which Part G, H, J, L or P of schedule 1 imposes a requirement".

Building work under ADJ – what you must do

- open-flue chimneys are defined as 'controlled service or fitting' under the Building Regulations. So as well as new chimneys having to meet the requirements of ADJ, existing chimneys must also meet these requirements if you carry out work on them
- if the work involves lining the chimney (either by introducing a new or replacement flue liner) this is defined as 'building work' under the regulations
- the liner (which may be of rigid, flexible or prefabricated components or may be cast in situ) could alter the flue dimensions, so you must test its performance, as with a new chimney system
- also, if a chimney is to be used for a different type of appliance, an appliance with a different output, or is being brought back into use, you must test it to ensure it is compliant

Compliance report – your responsibility

The person carrying out the work is responsible for meeting the requirements of ADJ. So when the building work is complete (e.g. a flexible metallic flue liner or chimney installed) you must notify the local BCB of the work undertaken.

The BCB needs to be notified about:

- any heat producing appliance you install or exchange (Building Regulations Part J); and
- associated fittings or services served by the appliance (Building Regulations Part L)

within a residential dwelling.

In relation to heat producing appliances covered by this manual, you are required to notify Gas Safe Register of the installation, who in turn will notify the relevant BCB on your behalf.

Figure 6.1 CORGI*direct* Chimney/hearth notice plate

CHIMNEY/HEARTH NOTICE PLATE

GAS safe REGISTER

Property address:	
Chimney/hearth installed in the:	
Is suitable for:	
Chimney liner (type/diameter):	
Suitable for condensing mode:	
Installed date:	
Installer name/Registration No.	

Ref. CP3PLATE

IMPORTANT SAFETY INFORMATION

THIS DATA PLATE MUST NOT BE REMOVED OR COVERED

Gas Safe is a registered trade mark of the HSE and is used under licence.

Notice plates

ADJ requires where a hearth, fireplace (including a flue box), chimney (new, extending or lining of an existing chimney system) is installed, essential information in the form of a notice plate containing relevant information on the correct application and use be permanently posted in the property.

Suggested locations for this notice plate are:

- next to the electricity consumer unit; or
- next to the chimney or hearth described; or
- next to the water supply stop-cock

See Figure 6.1 of an example of a typical notice plate produced by CORGI*direct* – Order Ref: CP3 Plate.

Additionally, ADJ recommends that a report be drawn up showing the materials and components used in the construction, and that the appropriate tests have been undertaken.

CORGI*direct* have for a number of years produced a suitable form for documenting the installation, which is designed in conjunction with the requirements of Appendix A of ADJ – see Figure 6.2 – Order Ref: CP3.

Figure 6.2 CORGI*direct* Chimney, fireplace and hearth commissioning record

To confirm the validity of the gas operative please contact the gas registration body.

Customer Ref:

CHIMNEY, FIREPLACE AND HEARTH COMMISSIONING RECORD

GAS safe REGISTER

This form complies with the requirements of relevant Building Regulations/Standards

Registered Business Details REG NO ____
Gas operative ____ (Print name)
Operative licence No. ____
Company ____
Address ____
Postcode ____ Tel No. ____

Building/Site Address
Name (Mr/Mrs/Miss/Ms) ____
Address ____
Postcode ____ Tel No. ____
Customer's signature ____

Client/Landlord Details (if different)
Name (Mr/Mrs/Miss/Ms) ____
Address ____
Postcode ____ Tel No. ____

General chimney, fireplace and hearth details		
Location of chimney, fireplace or hearth		
Application capability (gas/oil/solid fuel/all)		
Intended type of appliance (state type or make where possible)		
If open fire give finished fireplace dimensions		
Recommended ventilation requirements		
State type of ventilation and free area required		
Hearth construction (new or existing)		
Chimney inspection/testing		
Satisfactory visual inspection	Yes/No	
Chimney system swept	Yes/No	
Satisfactory coring ball check	Yes/No	
Satisfactory smoke test	Yes/No	
Satisfactory appliance spillage test ** When necessary*	Yes/No*	
Chimney, hearth notice plate fitted	Yes/No	
Chimney, hearth notice plate location	State	

Chimney construction	
Type/make	
New/existing	
Give internal flue dimensions (for natural draught appliances, provide the equivalent flue height where calculated)	
For clay/concrete chimney liners/blocks, confirm that all joints are socket end up and suitable jointing material is used	
For refurbished chimneys state type and make of liner used	
Details of chimney outlet including compliance criteria	
Details of any bends incorporated, together with their angles	
Details of provision for cleaning and frequency necessary	

I confirm that the work described in this record has been satisfactorily completed in accordance with Building Regulations/Standards, the current Gas Safety (Installation and Use) Regulations (GSIUR), appropriate standards and relevant manufacturer's/industry requirements.

Gas operative's signature **Date**

Key: Top Copy – Customer/client Green Copy – Gas Operative

To re-order quote Ref. CP3FORM

Installation

To avoid repetition, you will find general information in **Part 8 – General installation details – Wet central heating**, including details on boiler location, gas supply and ventilation.

General points you need to know

Gas fire/BBUs are manufactured, tested and approved for installation as a combined unit. The boiler is not designed to be installed alone without the fire, or the fire without the boiler.

Generally, gas-fired (NG or LPG) radiant convector, radiant convector fuel effect and Inset Live Fuel Effect (ILFE) fire/back boiler combinations are available – although its important to note that the latest high efficiency BBUs' now incorporate electric fires as opposed gas

- you may not carry out an installation 'marriage' using another manufacturer's appliance. But some manufacturers provide replacement gas fires that you can use to upgrade the fires of their earlier BBUs
- when you upgrade such an installation, you must make sure that the fire/BBU are compatible (check with the manufacturer)
- before you begin installing a combined gas fire/BBU see **Part 8 – General installation details – Wet central heating – Builder's opening**
- also verify the correct operation of the open-flued chimney system (see the current Essential Gas Safety – Domestic – Part 14 for further guidance)
- a chimney serving an open-flued gas fire/BBU must not serve another appliance - and you must route it to ensure full clearance of the POC safely to atmosphere (see the current Essential Gas Safety – Domestic – Part 14 for further guidance)
- installing a new or replacement open-flued fire/BBU to an existing chimney of proven performance - or to a newly installed chimney system - is no guarantee that it will operate correctly i.e. no spillage of POC when you test it
- if you are installing the gas fire/BBU onto an unlined brick/masonry chimney, line the chimney, using a flexible metallic flue liner
- connect all combined gas fire/BBU to the gas supply by a permanently fixed rigid pipe
- the final connection to the boiler must incorporate an isolating tap and means of disconnection to facilitate removal for servicing/maintenance etc.
- you normally install combined gas fire/BBU to an existing brick/masonry chimney and builder's opening
- if there is no existing opening or chimney, you can erect a false chimney breast to accommodate a flue box enclosure and chimney system (see **Part 8 – General installation details – Wet central heating – Chimney systems – Flue box/enclosure**)
- whichever chimney system you use, install the boiler and fire on a hearth, or sufficiently high on the wall - so as to prevent a fire hazard to carpets, furnishings etc.

Side wall and shelf protection requirements

Install the gas fire of the combined unit so that no part of a combustible wall (when measured laterally from the flame or incandescent radiant source) is less than 500mm, from that radiant source.

Note: The latest electrical fires used on HE BBUs may require smaller combustion clearances – consult the manufacturer's instructions.

Additionally, refer to the manufacturer's instructions for details of:

- limitations on the height and depth of any shelf above the appliance
- any protection needed to avoid the shelf reaching an excessive temperature

Air supply needed for safety

- for a combined open-flued gas fire/BBU to clear its POC it is essential there is adequate ventilation to the room in which it is installed
- calculate the ventilation requirements for a fire/BBU combination by adding the maximum heat input ratings of both appliances together in accordance with the manufacturer's instructions (see also the current British Standard for ventilation requirements: BS 5440-2 for further guidance)

Ventilation sizing, including routes and configurations are discussed in greater detail in the current Essential Gas Safety – Domestic – Part 4 – Ventilation.

Combined gas fire/back circulators – 7

Figures

Introduction

Combined gas fire/back circulator installations were a convenient method of using a space to locate a domestic hot water circulator. They were often installed in dwellings to replace an existing solid fuel back boiler.

- domestic gas circulators are designed for heating domestic hot water in a hot water storage vessel. As such, they are suitable for you to connect directly to the vessel
- be aware that some circulators have the capacity not only to heat the domestic hot water, but also a limited number of radiators. Connect these units using an indirect hot water storage vessel

In common with other forms of open-flued gas appliances, there are restrictions on suitable locations for safety reasons – see **Part 2 – Open-flued boilers** for guidance.

Installation

To avoid repetition, general information will be found in **Part 8 – General installation details – Wet central heating**, including details on boiler location, gas supply and ventilation.

General

- a back circulator is manufactured, tested and approved for installation together with a gas fire as a combined unit
- the back circulator must not be installed on its own. It must always be accompanied by a gas fire approved for such installation by the gas fire/back circulator manufacturer(s)
- when you upgrade such an installation with a replacement gas fire, you must check that the fire and back circulator are compatible. So you must check this information with the gas fire/back circulator manufacturer(s) – as only they may know
- before you begin installing a combined gas fire/back circulator see **Part 8 – General installation details – Wet central heating – Builder's opening**
- design and test the chimney to ensure correct operation (see the current Essential Gas Safety – Domestic – Parts 13 and 14 for further guidance)

A chimney serving an open-flued gas fire/back circulator must not serve any other appliance - and you should route it to ensure full clearance of the POC safely to atmosphere in accordance with the manufacturer's instructions (see also the current Essential Gas Safety – Domestic – Part 14 for further guidance).

Installing a new or replacement fire/back circulator unit to an existing chimney of proven performance, or newly installed chimney system is no guarantee that it will work correctly i.e. no spillage of POC when you test it.

- connect all combined gas fire/back circulators to the gas supply by a permanently fixed rigid pipe
- the final gas connection to the back circulator/fire must incorporate an isolating valve and means of disconnection to facilitate removal for servicing/maintenance, etc.

Combined gas fire/back circulators are normally installed to an existing brick/masonry chimney and builder's opening.

- in the absence of an existing opening or chimney, you can erect a false chimney breast to accommodate a flue box enclosure and chimney system (see **Part 8 – General installation details – Wet central heating – Chimney systems – Flue box/enclosure**)
- whichever chimney system you use, install the back circulator and fire on a hearth or sufficiently high on the wall to prevent a fire hazard to carpets, furnishings, etc.

Side wall and shelf protection, as well as ventilation provisions are as discussed in **Part 6 – Combined gas fire/back boilers**.

Unlined brick/masonry chimney

- gas fire/back circulators are normally designed for fitting to both unlined and lined brick/masonry chimneys. There is generally no provision for you to connect the back circulator flue outlet directly to any chimney lining
- when you plan installing a liner, refer to **Part 8 – General installation details – Wet central heating – Chimney systems – How to seal the annular space between the flexible metallic liner and the brick/masonry chimney** for how to connect it
- if you fit a gas fire/back circulator to an unlined chimney, the appliance flue outlet should:
    1. Prevent the entry of falling debris into the appliance flue spigot or flue connection piece; and
    2. Provide a void of a minimum volume of 12dm^3 below the lowest point of the appliance flue outlet (see Figure 7.1).
- where you install a gas fire/back circulator, you should not need to line the chimney provided the flue length doesn't exceed 10m (external wall) or 12m (internal wall)
- but if you install a liner, follow the manufacturer's instructions (see also **Part 8 – General installation details – Wet central heating – Chimney systems – Chimney liner – its function** for further guidance)
- for a gas fire/back circulator connected to an unlined brick/masonry chimney, you do not normally need to fit a flue terminal in place of, or to the existing chimney outlet

How to seal the fireplace opening

- you must use the gas fire manufacturer's closure plate when you fit a gas fire to a back circulator. But make sure there is access to the back circulator - and the facility to check draught diverter operation
- you may have to modify the closure plate and seal in position in accordance with the gas fire or back circulator manufacturer's instructions
- if you fail to follow these instructions - or to seal the closure plate effectively, this may lead to spillage of POC from the fire

The gas fire or back circulator manufacturer will give specific instructions on how you modify the gas fire closure plate and how you seal this to the back circulator framework and chimney breast.

When using a closure plate and an infill panel, if the fireplace opening is too large (see Note) secure the infill panel and seal it to the fireplace/wall opening or fire surround on all four sides, using an appropriate adhesive tape or other sealing method, e.g. fire cement and screws.

Make sure the tape or sealing material is suitable for the surface you affix it to. It must be able to maintain its seal up to a temperature of 100°C – proprietary tapes are specifically developed for this purpose i.e. tapes with the code PRS 10.

- to successfully seal to a tiled fireplace, ensure the surface is free from dust, polish, grease etc. - and is dry as well as clean
- on newly plastered walls or walls with porous surfaces, ensure that the plaster or surface is dry, free of dust, grease, etc. and a sealant applied and dried before you try to apply sealing tape
- if you apply tape to a surface without following these rules, the tape is almost certain to peel off. This will adversely affect flue performance and may cause spillage of POC into the room

Figure 7.1 Typical gas fire/back circulator installation

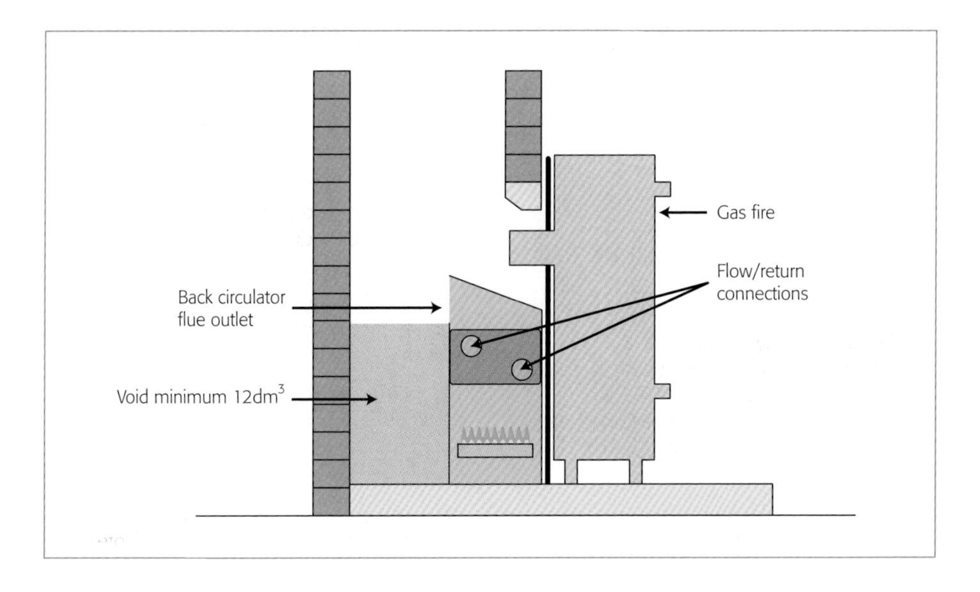

- also ensure that the closure plate is not bent or buckled, as this will place undue strain on the seal, which may ultimately fail
- tapes or sealants must be sufficiently flexible to seal along uneven surfaces such as rough stone fireplaces
- when you fit a gas fire/back circulator to a builder's/fireplace opening, do not fit combustible material inside this opening

Note: If the fireplace opening is too large to be closed off with the standard closure plate and back circulator framework, you can use an infill panel. The infill panel must be made from fire resisting material, having an opening to accommodate the back circulator framework and closure plate.

Secure the back circulator framework and closure plate to the infill panel - so that any debris that may fall down the flue can be removed.

Warning: The continued safe operation of the combined gas fire/back circulator depends on the closure plate and/or infill seal remaining effective until the appliances are next serviced.

Figure 7.2 Essential seal when fitting a commercial surround

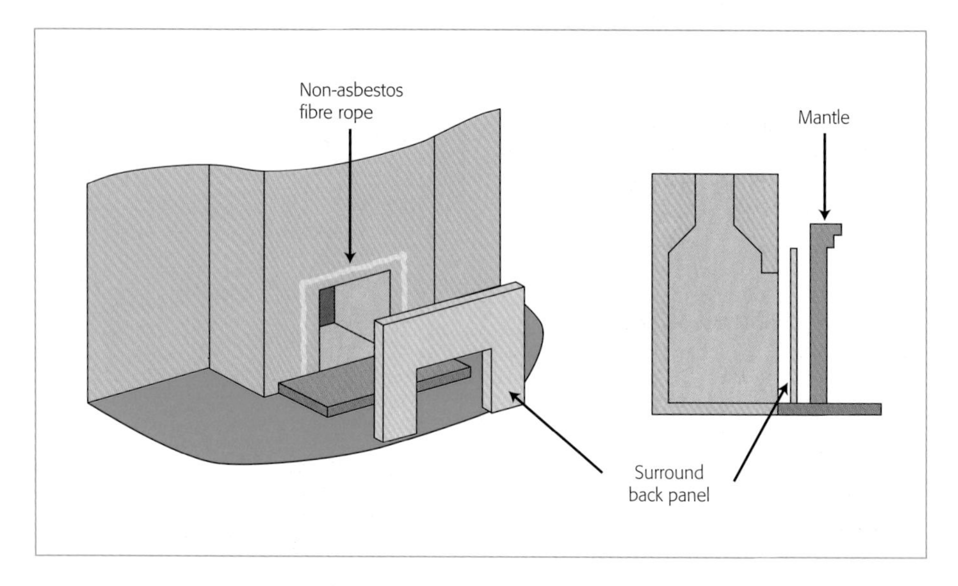

Commercially manufactured surrounds

- you will find the method of securing the surround to the wall in the manufacturer's installation instructions

It is important for the correct operation of the fire/back circulator and the clearance of the POC that there are no gaps between the surround and wall surface, or hearth that will admit air to the builder's opening and chimney (see Figure 7.2).

- where you use an infill panel in conjunction with the surround, it is equally important that you also seal this to the surround and wall surface to prevent air getting into the chimney
- from a gas safety point of view, there must be an airtight seal which stops air getting in, adversely affecting chimney performance (see **How to seal the fireplace opening** in this Part)
- equally, as it is within the fireplace opening, the sealing material should be non-combustible and sufficiently durable to continue to maintain a seal in normal circumstances

Note: When you secure a fire to a commercial surround incorporating a marble or similar infill panel, take great care to prevent cracking the marble - avoid bolts, screws and plugs that rely on expansion to secure a grip

How you secure the fire

- although a gas fire is designed to be free-standing when fitted to a back circulator, secure the fire to the back circulator framework in accordance with the manufacturer's instructions
- if you intend to wall mount the fire, use all the manufacturer's fixing holes
- in all cases it's essential you use screws of adequate length and size to secure the fire

- never 'secure' the fire using proprietary sealing compounds e.g. silicone sealant - unless the manufacturer's instructions specifically say so

Note: Whatever method you use for securing the fire to the wall, it should be robust enough for the fire to be removed for regular servicing, safety inspection etc.

Gas fire flue spigot restrictor: improves efficiency

Traditionally, gas fires are designed as stand alone appliances to be installed using a closure plate to seal off the fire grate opening. This concentrates the passage of air into the flue on the fire and whatever opening(s) provided by the fire manufacturer in the closure plate.

Gas fire manufacturers provide a flue spigot restrictor as they do not know:

- the type of chimney their fire will be installed to, or
- the amount of draught that will be created

The flue spigot restrictor is designed to slow down the passage of POC through the fire and thereby improve its overall efficiency.

- in the case of a fire fitted to a back circulator, make provision in the closure plate to allow air for combustion and for the draught diverter operation of the back circulator
- the back circulator manufacturer will require a substantial opening under the gas fire – where the closure plate would normally be – for this purpose
- as a result, air passing into the chimney is no longer concentrated on passing through the gas fire. So where a gas fire is fitted as a combined gas fire/back circulator, do not fit the restrictor to the gas fire spigot - unless specified by the appliance manufacturer

Circulator flow and return circuit

Direct hot water storage vessel

- where a back circulator is fitted to a direct hot water storage vessel all domestic hot water installation pipes must be non-ferrous
- a back circulator will often be connected to the existing gravity flow and return pipes and storage vessel connections. But to improve performance or if you use a new storage vessel, fit the water connections as close as practicable to the top of the vessel, using a mechanical joint
- the return pipe connection should be between 100mm and 200mm from the vessel base
- any mechanical joints you use to connect the flow and return pipes to a direct hot water storage vessel must comply with the current Water Supply (Water Fittings) Regulations
- the flow and return pipes from the back circulator to the storage vessel must be at least the same size as the heater connections, normally at least a minimum of 22mm copper pipe
- keep horizontal pipe runs as short as possible; the ratio of horizontal and vertical pipe should not exceed 4:1

Indirect hot water storage vessel

In known temporary hard water areas, only connect back circulators to indirect hot storage vessels, to reduce scale build-up.

General installation details – Wet central heating – 8

8 – General installation details – Wet central heating

8 – General installation details – Wet central heating

Figures

Tables

8 – General installation details – Wet central heating

Introduction

These are general guidance notes on installing new and replacement central heating boilers and fire/back boiler/circulators, common to most of the appliances covered in this Part.

Gas appliances in bathrooms – electrical zoning requirements

Although standards and codes of practice clearly document the electrical aspects of gas appliance installation, one area of continual debate is where to locate gas appliances - especially boilers in room(s) containing a fixed bath or shower.

For electrical safety reasons:

- only consider a bathroom or shower room if there is no alternative location
- with the vast range of flueing options now available, any need to use this location should be relatively rare. But you need the following information if you have no alternative

The electrical installation needs to comply with the requirements of BS 7671, which regulations introduce particular factors to be considered when you install in 'wet-room' locations.

Rooms containing fixed baths or showers have always been considered as a location where there is a higher risk of electrical shock.
This is due to:

  - body resistance when immersed in water or with wet skin
  - likely contact with substantial areas of the body and earth potential

- BS 7671 introduce a 'zonal' concept, where a series of zones are described which extend outwards from the source of highest risk. In most cases this will be the fixed bath, shower tray or cubicle

In essence, the bathroom/shower room is divided into 3 zones that each have specific dimensions and are designated 0 – 2 (see Figure 8.1 for zone designations).

Electrical zones

The following zones describe the restrictions placed on electrical installations/equipment installed in rooms containing a fixed bath or shower.

Zone 0

This is the interior of the bath tub or shower basin or, in the case of a shower area without a tray, it is the space having a depth of 100mm above the floor out to a radius of 1200mm from a fixed shower head.

Points to note:

- only SELV not exceeding 12V d.c. rms or 30V ripple-free d.c. may be used as a protective measure against electric shock, the safety source being outside Zones 0, 1 and 2
- no switchgear or accessories are permitted
- current using equipment, which is fixed and permanently connected is permitted provided it is specifically designed for use in this zone
- equipment designed for use in this zone must be to at least IP X7 (immersion)
- wiring associated with equipment in this zone may be installed, either on the surface or at a minimum depth of 50mm on walls limiting these zones. For Zone 0 this would only be applicable to shower areas without basins

Figure 8.1 Bathroom zones

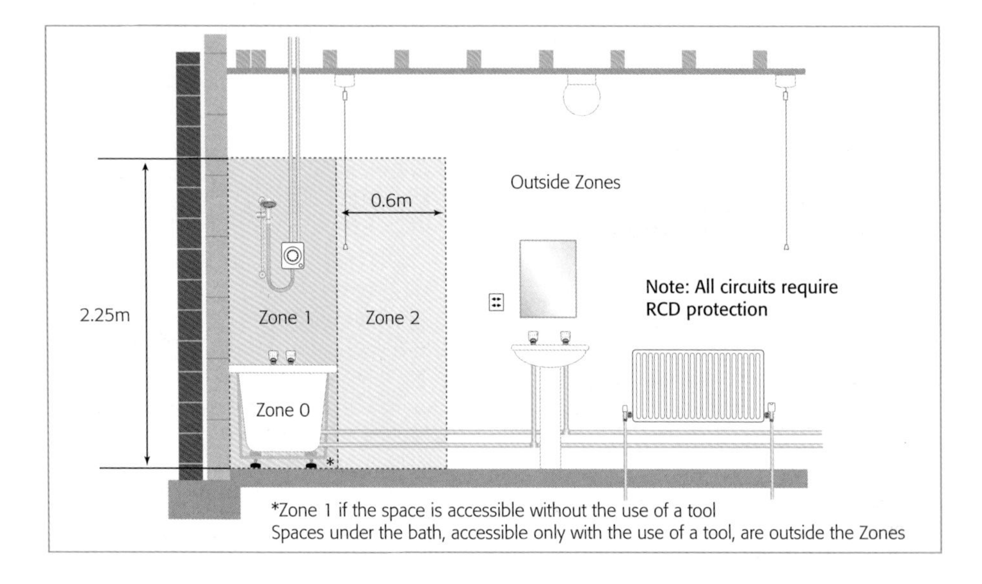

Zone 1

This extends above Zone 0 around the perimeter of the bath or shower basin to 2.25m above the floor level, and includes any space below the bath or basin that is accessible without the use of a key or tool.

For showers without basins, Zone 1 extends out to a radius of 1200mm from the centre point of a fixed showerhead.

Points to note:

- other than switches and controls of equipment specifically designed for use in this zone and insulating cords or cord operated switches, only switches of SELV circuits are permitted
- provided they are suitable, fixed and permanently connected items of current using equipment may be installed in this zone. Such items are whirlpool units, electric showers, shower pumps, SELV and PELV protected equipment, ventilation equipment, towel rails, water heating appliances and luminaires
- equipment designed for use in this zone must be to at least IP X4 (splashproof), or IP X5 (jet proof) where water jets are likely to be used for cleaning purposes
- wiring associated with equipment in this zone may be installed, either on the surface or at a minimum depth of 50mm on walls limiting this zones

Zone 2

This extends 600mm beyond Zone 1 and to a height of 2.25m above floor level.

Points to note:

- other than switches and controls of equipment specifically designed for use in this zone and insulating cords or cord operated switches, only switches of SELV circuits and shaver units to BS EN 61558-2-5 are permitted
- there is no restriction to current using equipment for use in this zone
- equipment for use in this zone must be to at least IP X4, or IP X5 where water jets are likely to be used for cleaning purposes
- wiring associated with equipment in this zone may be installed, either on the surface or at a minimum depth of 50mm on walls limiting this zones

Socket outlets, other than SELV sockets and shaver units to BS EN 61558-2-5, are not permitted within a distance of 3m from the boundary of Zone 1.

Additional protection (RCDs)

In addition to the normal protection against shock by automatic disconnection of supply, all circuits in the location must be protected by 30mA RCDs.

Builder's opening

- if you install a back boiler/circulator in a fireplace opening, it must have sufficient space around it for air to circulate and for draught diverter operation
- the opening must be large enough to accommodate the appliance and comply with manufacturers specified dimensions
- where the existing flow and return pipes are sited on the opposite side to the boiler connections, the pipes must not cross in front of the boiler or heat exchanger access opening in a way that would impair access for servicing
- where the boiler manufacturer provides a left or right handed boiler, reverse the heat exchanger to eliminate this problem
- to assist the correct operation of the fire/back boiler/circulator, the builder's opening or enclosure should have only two openings: an entry through and around the rear of the fire and an exit via the chimney
- seal all other openings, gaps and cracks and pay special attention to those between any surround and the builder's opening, those which exist in respect of an existing under floor air supply and those made for the passage of gas, water, flue pipes and electric cables
- the material you use for sealing must be fire-resistant or non-combustible if you use it within the fireplace opening

Note: Where you need to provide ventilation, do not install the air vent directly within the builder's opening or fireplace recess (see 'Flame reversal' in this Part).

How to protect pipework

- protect water and gas pipework that you install within the builder's opening from potential corrosion and damage caused by soot and debris that might fall from the chimney
- a suitable method of protection is to wrap the pipe with a suitable tape, e.g. PVC tape
- you must sleeve all gas pipes where they pass through solid walls (see the current Essential Gas Safety – Domestic – Part 5 for further guidance)

Hearth

Where a hearth is required within the builder's opening for a fire/back boiler/back circulator installation, build a constructional hearth:

1. Construct it of solid, non-combustible material, at least:
   a) 125mm thick; or
   b) 25mm thick and placed on non-combustible supports at least 25mm high.
2. It must be able to support the total weight of the back boiler/back circulator and fire.
3. It should extend not less than 150mm from the back and sides of the boiler/back circulator. If there is a wall within 150mm of the back boiler/back circulator, the hearth should extend to that wall. In all installations the hearth should extend to the front of the back boiler or back circulator.

Secure the gas fire/back boiler/back circulator to the hearth using the manufacturer's securing method, provide a hearth for the fire unless the fire is to be wall-mounted.

Where a hearth is required, it must comply with the following:

- It must be made from fire-resisting material
- It must be a minimum thickness of 12mm
- It must extend 300mm forward from the back plane of the gas fire
- It must extend at least 150mm beyond each edge of the naked flame or incandescent radiant source

Where you intend to wall-mount the fire, any flame or incandescent material must be at least 225mm above any carpet or floor covering. Where the floor is likely to be covered, any flame or incandescent materials needs to be at least 300mm above the floor in order to make allowances for floor coverings beneath the fire.

Note: In the case of an inset live fuel effect (ILFE) gas fire/back boiler installation, an upstanding edge of 50mm minimum height along the front and sides of the hearth or the installation of a fender 50mm high or more, discourages carpets or rugs being placed on top of the hearth.

Brick/masonry chimneys – factors to be aware of

Most gas fire/back boiler/back circulator installations are installed using the existing builder's opening and brick/masonry chimney (which were designed for solid fuel appliances).

As a result, most chimneys have a deposit of soot (containing particles of sulphur) on the inner walls of the chimney throughout its length. When water (rain) mixes with sulphur, sulphuric and nitric acids are formed. These attack the mortar joints and brickwork of the chimney - and ultimately cause the chimney stack to lean to one side before eventually falling over.

Condensation within the chimney contributes to this problem. It forms when water-laden air or, POC come into contact with a cold surface.

In the case of a gas appliance, the POC produced by the appliance contain water vapour. Typically a 30kW (gross) boiler operating for 1 hour will produce approximately 4 litres of water in vapour form during this period. A 6kW gas fire combined with a 6kW back circulator will produce just under 1.5 litres in the same time period.

Unless steps are taken to prevent this water vapour from condensing onto the cold chimney, condensation will form (see the current Essential Gas Safety – Domestic – Part 13 for further guidance).

How to help prevent condensation

- the POC should be kept warm and insulated from the cold brick/masonry chimney
- one way you can achieve this is to insert a flexible metallic flue liner into the chimney, which is then connected directly to the back boiler using the boiler manufacturer's approved method. In the case of gas fire/back circulator installations see **Part 7 – Combined gas fire/back circulators – Unlined brick/masonry chimney**
- thoroughly sweep a chimney used for a fuel other than gas, before you install the flexible metallic flue liner (see **Chimney liner – its function** in this Part)
- if you install a gas fire/back boiler into an unlined brick/masonry chimney, line the chimney with a flexible metallic flue liner

Note: Some chimney stacks are designed to be a feature of the dwelling and as such, their size and shape may be larger than normal (i.e. the chimney top may be in excess of 1000mm x 1000mm serving a single chimney).

If so, the chimney top may take on the dimensions of a flat roof and create wind turbulence around the terminal, interfering with the evacuation of the POC (creating intermittent down draught).

If this happens, you may need to raise the terminal by a minimum of 250mm above the 'flat roof effect' created by the chimney stack.

Boiler locations

Compartment installation requirements

A compartment is an enclosure specifically designed or adapted to house a gas appliance (see Figure 8.2).

Open-flued or room-sealed boilers installed in compartments must comply with these requirements:

1. The compartment must be a fixed rigid structure.
2. If the appliance manufacturer's installation instructions give no specific advice, any internal surface of the compartment which is constructed of combustible material, must be at least 75mm away from any part of the appliance. Alternatively line the surface with non-combustible material with a fire resistance of not less than 30 minutes. Materials that comply with the relevant part of BS 476 meet this requirement.
3. The compartment must incorporate air vents for the provision of air for compartment cooling (if applicable) and where necessary, combustion and correct operation of an open-flued chimney system in accordance with the manufacturer's instructions (see also the current British Standard for ventilation requirements: BS 5440-2 and the current Essential Gas Safety – Domestic – Part 4 for further guidance).
4. If the compartment houses an open-flued boiler, the door or air vents must not communicate with a bath/shower room. If the air vents communicate with a bedroom/bedsitting room, you must install according to the guidelines in **Part 2 – Open-flue boilers – Restricted locations – for safety reasons – Bedroom/bedsitting rooms**.

Figure 8.2 Compartment installation

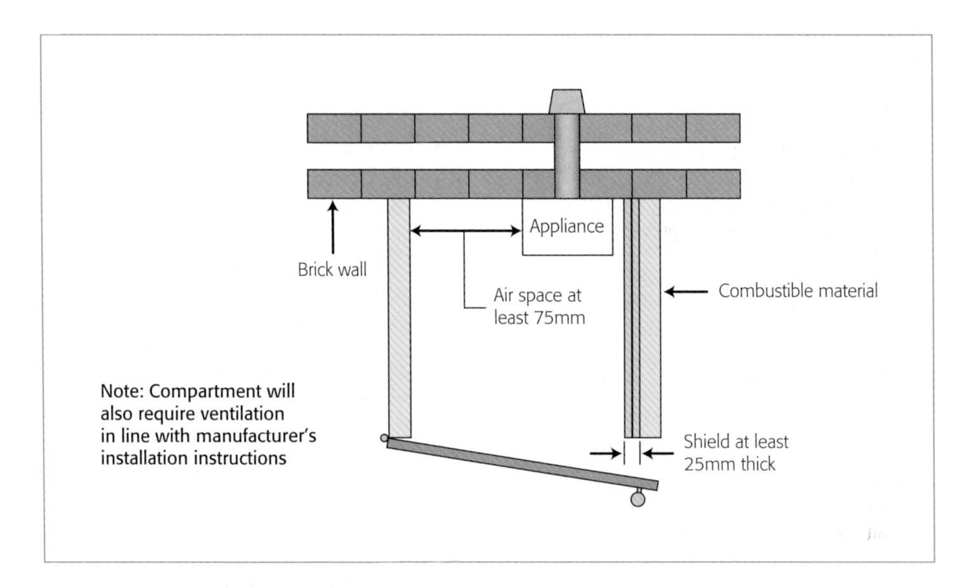

5. The compartment must permit access for inspection and servicing of the appliance and any ancillary equipment. To discourage its use as a storage cupboard, fix a notice in a prominent position to warn against such use. It must be fitted with a door that will permit withdrawal of the appliance and any ancillary equipment.

Attention: For compartments or cupboards located on an upper floor and containing an open-flued appliance – take care to ensure this location doesn't interfere with the safe operation of any open-flued appliance sited in a downstairs location.

Be aware that under certain adverse weather conditions and where there are poorly designed chimneys, open-flues may fail to operate safely and spill POC (see the current Essential Gas Safety – Domestic – Part 14 for further guidance).

Airing cupboard installations – how to avoid problems

An airing cupboard is generally sited centrally within a dwelling, often on an upper floor. Previously, when a central heating boiler had to be installed in one, the appliance flueing option was limited to the open-flued type.

Modern room-sealed boilers are fanned draught room-sealed and have a variety of flueing options. This prevents the previously often-found problem of lint (from clothes) causing premature blockage of the burner primary airways. This blockage can lead to:

- deteriorating combustion
- blockage of the heat exchanger with soot; and
- the inherent risk of fire that has always been a real problem for open-flued boilers in this location, even when the boiler is separated from the airing space

In addition, siting an open-flued boiler in an upstairs location needs combustion and compartment ventilation. As a result, the dwelling is now flued and ventilated to atmosphere at high level.

- in buildings of tight construction (such as timber frame dwellings or where the dwelling has been draught-proofed with double glazing) this may prevent the chimney of any open-flued appliance operating correctly - particularly if it is sited in a downstairs location.
- the chimney serving the downstairs appliance may be subject to thermal inversion, and the appliance subject to spillage of POC into the room (see **Compartment installation requirements** in this Part). Also see the current Essential Gas Safety – Domestic – Part 14 for further guidance
- wherever possible, consider installing room-sealed or fanned draught room-sealed boilers in airing cupboards
- where you install a boiler in an airing cupboard, ensure it complies with the requirements for compartment installations
- where combustion air or ventilation is required, install in accordance with the manufacturer's instructions

See also the current British Standard for ventilation requirements: BS 5440-2 for further guidance.

- additionally, separate the airing space from the boiler compartment by a non-combustible partition (perforated by apertures with minor dimensions no greater than 13mm)

Expanded metal or rigid wire mesh are suitable materials for this partition.

- where the boiler is of the open-flued type, the draught diverter and the air vents should be in the boiler compartment
- unless surrounded by an air inlet duct, no chimney must pass through the airing space - unless the chimney is protected sufficiently to prevent damage to the airing space contents

Double-walled flue pipe conforming to BS EN 1856-1 may satisfy the requirements for a 25mm air gap from combustible materials.

- ensure that single-walled flue pipe up to 1m from the draught diverter outlet connection is protected by an air gap of at least 25mm. You can provide this air gap by a non-combustible guard that forms an annular space around the chimney of not less than 25mm

Expanded metal or rigid wire mesh are both suitable materials for the guard.

External installations – factors for you to consider

If you install a boiler externally, it needs to be either:

- Specifically specified by the manufacturer as suitable for external installation without the need for additional protection, or
- Installed in an enclosure that can provide permanent weather protection.

If an enclosure is needed, it needs to comply with the compartment installations requirements and follow the appropriate Building Regulations:

- Within the enclosure, there needs to be an accessible waterproof means of electrical isolation of the boiler installation (see **Electrical connections – your responsibility** in this Part).
- Fit the enclosure with air vents, where required, direct to outside air at high and low level.

Note: When providing low level ventilation the vent must not be less than 300mm above ground level.

- any permanent openings in the enclosure, including those in the air vents, need to have a minor dimension not greater than 16mm – in order to prevent birds or rodents getting in
- this dimension should not be less than 6mm – so as to minimise the risk of blockage
- if you locate a boiler in an exposed position (such as an external location, roof space or unheated garage etc.) provide adequate insulation/frost protection
- if you fit a frost thermostat, it is recommended you adjust it to operate at 4°C

Roof space installations – what you need to do

Roof spaces incorporating boiler installations needs to comply with the following requirements:

- provide vertical clearances to meet the static head requirements of open vented systems.
- provide flooring area for normal use and servicing under and around the boiler (see Note 1).
- provide a permanent means of access to the boiler installation (see Note 2).
- provide fixed lighting for the boiler installation and access.
- provide a guard to prevent contact between stored articles and the boiler installation.
- provide adequate insulation/frost protection

Note 1: When you install an open-flued gas boiler, any provision required for the protection of the floor or wall on which you are mounting the boiler will be detailed in the boiler manufacturer's installation instructions.

- in the absence of instructions, provide a non-combustible insulating base of at least 12mm thickness under the boiler – if the floor supporting the boiler is of combustible material
- in the case of a floor-standing boiler, make sure that the floor on which you place the boiler can support its weight
- also ensure that, should the floor be exposed to a prolonged period of wetness, (due to a water leak for example) its strength would not be impaired (e.g. chipboard flooring supporting a boiler under these circumstances may collapse under the weight of the boiler)
- equally, in the case of a wall-mounted boiler, ensure that the wall can support its weight - and you use the correct number, size and length of fixing screws or bolts recommended by the manufacturer

Note 2: A permanently fixed retractable roof ladder is considered to satisfy the requirement for a purpose-designed means of access. Provide a safety guard around the roof access opening.

Where the boiler is open-flued, provide ventilation and flue the boiler in accordance with the manufacturer's instructions.

See the current British Standard for ventilation requirements: BS 5440-2 also the current Essential Gas Safety – Domestic – Part 4 for further guidance.

- where you use an existing brick/masonry chimney that is not lined, seal the unused lower portion of the chimney from the used portion by means of a plate approximately 250mm below the appliance connection to the chimney. This will provide a catchment area/void for debris collection
- also provide an access panel in this area, to enable inspection and clearance of any debris

- ensure that other chimneys in the same chimney stack are not sealed off

Note: Permanently close off any openings into the lower portion of an internal chimney i.e. the sealed-off section.

For a chimney with at least one external face on the outside wall, ventilate the sealed section to the external air at high and low level to prevent damp penetration.

Under-stairs cupboards

A boiler you install in an under-stairs cupboard must comply with one of the following:

- Where the premises in which the cupboard is located is no more than two storeys, the cupboard needs to comply with the requirements for compartment installations.
- Where the premises in which the cupboard is located is more than two storeys, all the internal surfaces of the cupboard, including the base, should be non-combustible - or should be lined with non-combustible material with a fire resistance of not less than 30 minutes. Materials that comply with the relevant parts of BS 476 will meet this requirement. The air vents should be direct to outside air. Size them in accordance with the manufacturer's instructions.

See also the current British Standard for ventilation requirements: BS 5440-2 for further guidance.

Note: Stairs are very often the only means of escape in the event of a fire: so only consider this location when there is no other practicable position.

Why clearances needed around the boiler

When you install the boiler, refer to the manufacturer's instructions for clearances around it. These clearances are to ensure sufficient air circulation for draught diverter operation, fireproofing (where necessary), servicing and maintenance (see **Boiler locations** in this Part).

Note: Some manufacturers recommend clearances in excess of 100mm, but claim that the boiler can be serviced with only a 25mm clearance. It is important you note that the additional clearance is needed to enable changing major components on the boiler without needing to remove the boiler from the wall.

Always follow the manufacturer's required installation clearances.

Commissioning: your responsibility

You are responsible for ensuring:

- all work has been carried out in accordance with the relevant Regulations (see **Introduction** at the beginning of this Part)
- that the gas appliance and installation operate in a safe and satisfactory manner
- when you connect a gas supply to an appliance, the Gas Safety (Installation and Use) Regulations require you to commission the appliance
- unless you can complete this work immediately, you must disconnect the appliance from the gas supply and label it accordingly
- you must test all gas fittings forming part of the installation for gas tightness - and purge of air (see the current Essential Gas Safety – Domestic Parts 6 & 15 for further guidance)

You will find additional information on LPG tightness testing in the Gas Installer Manual Series – Domestic – 'LPG – Including Permanent Dwellings, Leisure accommodation Vehicles, Residential Park Homes and Boats'.

Follow the manufacturer's commissioning instructions supplied with the appliance.

You may use the following general procedure for appliances covered in this Part. You will find additional requirements specific to individual appliances addressed separately afterwards.

General procedure

1. Check that the ventilation requirements are correct and in accordance with the manufacturer's instructions. See also the current British Standard for ventilation requirements: BS 5440-2 for further guidance.
2. Check that the chimney termination is correct (see the current Essential Gas Safety – Domestic – Part 13 for further guidance).
3. Where applicable, carry out a flue flow test (see the current Essential Gas Safety – Domestic – Part 14 for further guidance).
4. Check electrical connections (see **Electrical connections – you are responsible for compliance** in this Part).
5. Test all appliance gas connections with non-corrosive leak detection fluid (LDF).
6. Check for correct operation of all control valves and that the ignition system(s) operate(s) correctly.
7. If necessary, adjust the pilot flame, to envelop the thermocouple tip. Ensure that it maintains the flame supervision device (FSD) correctly.
8. If the pilot light is extinguished, do not try to re-light it for 3 minutes. Check the 'fail safe' operation of the FSD according to the manufacturer's instructions (see the current Essential Gas Safety – Domestic – Part 12 for further guidance).
9. Check the operating pressure(s), gas rate(s) or both are in accordance with the appliance data plate. Adjust as necessary (see the current Essential Gas Safety – Domestic – Part 11 for further guidance).
10. Check all the burners cross light - and the flame picture is satisfactory in terms of stability, structure and colour.
11. Where applicable, carry out a spillage test (see the current Essential Gas Safety – Domestic – Part 14 for further guidance).
12. Check the boiler thermostat is operating correctly.
13. Flush water system (see **System requirements** in this Part).
14. If it is a sealed system, check the pressure in the pressure vessel is correct and that the system pressure is adequate (see **System requirements – Sealed system** in this Part).
15. Check the boiler/system bypass valve is correctly adjusted (see **System bypass – adjust if needed** in this Part).
16. Balance the system (see **How to balance the system** in this Part).
17. Ensure any compartment warning labels are correctly fixed (see the current Essential Gas Safety – Domestic – Part 10 for further guidance).
18. Instruct the user on how to operate the appliance(s) and controls.
19. Complete all relevant documentation – Benchmark for example – and leave all instructions with the gas user/responsible person.

20. Advise the user that the appliance(s) requires servicing/safety checks at a minimum of 12 monthly intervals - or at intervals specified in the manufacturer's instructions.
21. Advise the user of any appliance/installation defects in writing. Where necessary, follow the current Gas Industry Unsafe Situations Procedure (see also the current Essential Gas Safety – Domestic – Parts 8 and 10 for further guidance).

Note: A gas appliance in normal use needs servicing and a check for safety at 12 monthly intervals, but this period depends on the amount of use and the type of room or space it is installed in. So it may require servicing at intervals less than 12 months.

Checks to make for combination boilers:

For a combination boiler to perform to the manufacturer's specification, you need to check and adjust the domestic hot water flow rate through the appliance. Manufacturer's instructions generally specify a preferred water flow rate in litres per minute raised by °C.

- check this performance by using a weir gauge (or suitable measuring jug) and thermometer when commissioning
- always use a suitable thermometer, either electronic or mercury glass file type. Place the thermometer in the water flow from a draw-off point and leave until the heater has reached its maximum temperature
- adjust the water flow to the heater as necessary against the manufacturer's specification

Checks to make for combined gas fire/back boiler/back circulators:

- ensure that the builder's opening has been correctly constructed and sealed in accordance with the manufacturer's instructions (see also **Boiler locations** and **Builder's opening** in this Part)
- commission the fire in accordance with manufacturer's instructions. The general procedure guidelines are detailed in the current Gas Installer Manual Series – Domestic – 'Gas Fires and Space Heaters'

Checks to make for condensing boilers:

1. Check that the flow/return temperature differential is 21°C.
2. Check that the syphon is clear and unobstructed and that the condensate discharge termination is in accordance with the manufacturer's instructions (see also **Part 5 – Condensing boilers** for further guidance).

System bypass – adjust if needed

Some boilers, in particular low water content boilers, incorporate or require a system bypass controlled by a valve. The purpose of the bypass is to prevent water overheating and boiling causing noise (kettling) in the boiler.

To avoid kettling, adjust the bypass valve so that the flow of water is always above the minimum required to prevent overheating of the boiler. The bypass is usually a minimum of 15mm size and positioned after the circulating pump and installed/adjusted in accordance with the appliance manufacturer's instructions.

How to balance the system

Due to the inevitable restrictions caused by pipework and fittings, water circulation through a central heating system will try to take the easiest route.

- so you must balance the system to ensure the distribution of hot water reaches all parts. It is important to achieve a temperature difference of 11°C for normal heating systems - and 21°C temperature difference for condensing boiler installations measured across the flow and return connections at the boiler
- you will also need to measure the temperature difference across individual radiators, the hot water storage vessel and (where fitted) the system bypass
- the circulating pump speed setting will also affect the balancing process, so adjust it in accordance with the circulating pump manufacturer's instructions

Servicing

The general procedure outlined here is for information purposes only. It is applicable to most boilers however, modern condensing boilers work to tighter tolerances and do not necessarily allow, or require a full strip and clean process.

General procedure (for all appliances when applicable)

Always follow the manufacturer's servicing procedure. In the absence of instructions, you may follow the general procedure for appliances covered in this manual

Where there are additional requirements specific to a particular appliance, these are addressed separately after this procedure

Preliminary examination

1. Check with the customer to ascertain any problems with the appliance and/or heating system.
2. Check the location of the appliance is suitable (see **Restricted locations** in the relevant Part of this manual).
3. Check for any damage that exists on the appliance and surroundings and advise the customer where appropriate before you start any work.
4. Check the operation of the appliance controls, including thermostats, ignition systems and flame supervision devices.
5. Check the appliance burner flame picture(s) and/or, measure the combustion process using an Electronic Combustion Gas Analyser (ECGA) to ascertain the extent of servicing required.

Note: Gas operatives need to be competent in the use of ECGAs', otherwise the results obtained could be misinterpreted leading to an unsafe gas appliance.

For detailed guidance on using ECGAs' refer to CORGI*directs* Gas Installer Manual Series – Domestic – 'Using portable electronic combustion gas analysers – Servicing and maintenance Domestic'.

6. Where applicable, check that the electrical installation complies with the Requirements for Electrical Installations – BS 7671.
7. Check clearances from combustible materials e.g. compartments etc.
8. Check the gas installation pipework for exposure to corrosion/sleeving and clearances from electrical cables.

Full service

1. Isolate the appliance from the gas and water supplies and where applicable, the electricity supply (see **Electrical connections – you are responsible for compliance** in this Part).
2. Because of the possibility of stray electrical currents, consider attaching a temporary continuity bond to the gas supply and the appliance (see the current Essential Gas Safety – Domestic – Part 5 for further guidance).
3. Remove the main burner for cleaning. Wherever possible dismantle the burner and remove any internal filter or lint arrester gauze. Clean as follows:
   a) Remove all surface dust with a paint brush or similar.
   b) Using a combination of brushes, remove dust and lint from within the primary air ports, venturi and burner(s).
   c) Check the burner(s) for cracks and metal fatigue.
   d) Investigate the cause of any faults found in c) above and correct. If the gas fire burner is affected (see **Flame reversal** in this Part).
4. Clean the main burner injector(s).
5. Remove the pilot assembly – clean the burner and injector.
6. Check the pilot supply tube is clean and unobstructed.
7. Reassemble the burner(s) and pilot assembly.
8. Check the condition of ignition leads and alignment of the electrode.
9. Access the heat exchanger and thoroughly clean it using a suitable flue brush or tool (see **Heat exchangers: danger if flueways not clean** in this Part).
10. Where applicable, examine and, where necessary, clean any fan associated with the appliance or flue.
11. Refit the burner assemblies and check all seals as necessary.
12. Test all disturbed joints for gas tightness using non-corrosive LDF
13. If room-sealed, check that the appliance case seals are in good condition; renew any sealing material as necessary. Also ensure that the case itself fits securely and that all fixing bolts/screws are located correctly (see also **Testing appliances with positive pressure cases – steps to help you ensure safety** in this Part).
14. Restore the electrical supply.
15. If necessary, adjust the pilot flame, to envelop the thermocouple tip. If the pilot light is extinguished, do not try to re-light the appliance for at least 3 minutes.
16. Test the FSD for correct operation (see the current Essential Gas Safety – Domestic – Part 12 for further guidance).
17. Re-light and check the appliance gas pressure(s), gas rate(s) or both in accordance with the appliance data plate and adjust as necessary.
18. Check the main burner(s) and pilot for satisfactory flame picture and/or, measure the combustion process using an ECGA – see Note to 'Preliminary examination'.
19. Check the ventilation requirements are correct and in accordance with manufacturer's instructions. See also the current British Standard for ventilation requirements: BS 5440-2 for further guidance.
20. Where applicable, carry out a flue flow and spillage test and check that the flue termination is correct (see the current Essential Gas Safety – Domestic – Parts 13 and 14 for further guidance).

21. Where applicable, check that the room-sealed terminal is installed correctly (fit a terminal guard where necessary) and that no undergrowth will interfere with combustion and adversely affect flue performance.
22. Check boiler/system bypass valve is correctly adjusted (see **System bypass** in this Part).
23. If it is a sealed system, check that the pressure in the pressure vessel is correct and that the system pressure is adequate (see **System requirements – Sealed system** in this Part).
24. Advise the user to have the appliance(s) serviced/safety checked at a minimum of 12 monthly intervals, or at intervals specified in the manufacturer's instructions.
25. Advise the user of any appliance/installation defects in writing and where necessary, follow the current Gas Industry Unsafe Situations Procedure (see also the current Essential Gas Safety – Domestic – Parts 8 and 10)

Attention: Whilst there are no specific instructions for servicing a multifunctional gas control valve, you should check the control knob is free and easy to operate when depressing the pilot control knob to establish the pilot flame.

- **if this is not the case, remove the plastic knob and apply a small amount of the control manufacturer's lubricating oil to the spindle**
- **failure to correct this fault could lead to a serious gas escape on the control**
- where gas appliances are fitted with an atmosphere-sensing device (ASD), service these devices strictly in accordance with the manufacturer's instructions. They are not 'field adjustable'. This means that if a fault develops on, for example, the thermocouple lead, you may need to replace the entire unit
- some manufacturers recommend replacing the ASD every five years (see the current Essential Gas Safety – Domestic – Part 12 for further guidance)

Note: Check gas boilers connected to a fanned draught chimney system to ensure that the burner will shut down in the event of failure of the draught (refer to manufacturer's instructions).

Where any room or premises is fitted with a fan (e.g. decorative re-circulatory ceiling fan, an extract fan, or a fan incorporated within an appliance for example a tumble dryer), the operation of the fan(s) must not adversely affect the performance of the chimney when you test the appliance in accordance with the manufacturer's instructions (see the current Essential Gas Safety – Domestic – Part 14 for further guidance).

Combination boilers

See **Commissioning: your responsibility** in this Part.

Combined gas fire/back boiler/ back circulators

- ensure the builder's opening has been correctly constructed and sealed in accordance with the manufacturer's instructions (see also **Boiler locations** and **Builder's opening** in this Part for further guidance)
- examine the gas fire heat exchanger (BBU/back circulator installation) especially at the rear of the radiants, logs or coals locations for cracks and general metal fatigue (see **Flame reversal** and **Commissioning: your responsibility** in this Part)

Condensing boilers

See **Commissioning: your responsibility** in this Part.

Heat exchangers: danger if flueways not clean

Some boiler heat exchangers have flue ways that are both horizontal and vertical whilst others have flue ways that are difficult to reach and clean.

- when you service such boilers you need to ensure all flue ways are clean, particularly if the heat exchanger is full of soot
- if you fail to clean these flue ways effectively, it will restrict or prevent the POC from passing through the heat exchanger. This will almost certainly result in soot blocking the boiler again
- during this period the user may be in danger from carbon monoxide (CO) poisoning, especially from open-flued appliances

Chimney flue blocks and liners: also keep clean

A chimney flue block system or liner serving a gas appliance may, over time, become restricted by spiders' webs. These may cause spillage of the POC into the room. If you suspect this, clean the chimney along its entire length.

System bypass

See **Commissioning: your responsibility** in this Part.

Central heating system checks

Where a sealed system appliance is installed, check:

- the water pressure is sufficient
- all pressure and temperature safety valves operate correctly and discharge to a safe position

See **System requirement** in this Part.

Maintenance

If you carry out any maintenance work on a gas appliance (e.g. clearing a blocked pilot jet etc.) the Gas Safety (Installation and Use) Regulations require you to examine:

1. The effectiveness of any flue.
2. The supply of combustion air.
3. Its operating pressure or heat input or, where necessary, both.
4. Its operation so as to ensure its safe functioning.

You must then take all reasonable practicable steps to notify any defect to the responsible person and where different, the owner of the premises in which the appliance is situated.

If neither is reasonably practicable, in the case of an appliance supplied with LPG, you must notify the supplier of the gas to the appliance, or, in any case, the transporter.

Fault finding – you must be methodical

The operation of appliance, burners, control taps, ignition systems, thermostatic controls and flame supervision devices are covered in detail in the appropriate parts of the current Essential Gas Safety – Domestic – Parts 1 – 17. However, the following list helps you take a methodical approach to fault finding.

General fault finding guide

1. Check with the customer to ascertain what particular problems they have had with the appliance. This helps to pin point any defects.
2. Check the location and general installation requirements for the appliance comply with the manufacturer's installation instructions.

3. Where possible, always refer to the appliance manufacturer's installation/maintenance instructions. They often contain fault finding information including flow charts, to guide you to a satisfactory conclusion

Tables 8.1 and 8.2, list a number of faults and possible causes relating to central heating systems and appliances.

Table 8.1 Fault finding chart – Systems

Symptom	Possible cause	Remedy
Pumped central heating (C/H) – domestic hot water (DHW): Upstairs radiators hot.	No anti gravity valve fitted or, existing valve not seating correctly.	Fit new anti gravity valve in C/H flow circuit or maintain existing valve.
Reduced or no circulation of hot water to hot water storage vessel (HWSV).	No water in circulating pipework.	Check water level/operation of ball valve in feed and expansion cistern.
	Cold feed blocked between feed and expansion cistern and system pipework. NB. Cistern may appear to be full.	Clear blockage.
	Old cylinder thermostat temperature control valve on HWSV return connection seized in closed position.	Maintain or replace valve.
	Air lock in circulation pipes.	Remove air paying particular attention to long horizontal pipe runs under floors. Circulating pipework should rise continuously from boiler to HWSV.
	Insufficient circulation head. NB. There should be a minimum of 1m between the centre of the boiler casting and the HWSV coil/calorifier.	Lower boiler or raise HWSV to obtain minimum circulation head.
Indirect single feed HWSV: Discoloured domestic hot water.	HWSV internal expansion vessel too small to accept – expanded system water (air seal(s) have been lost).	Replace HWSV with correctly sized internal expansion vessel.
Pumped systems general: Water flowing from warning pipe serving C/H feed and expansion cistern.	HWSV (direct type) immersion calorifier leaking, allowing water to flow from the domestic cistern to the C/H feed and expansion cistern via the defective immersion calorifier.	Replace immersion calorifier or exchange direct HWSV for indirect type and appropriate sized feed and expansion cistern.
	Indirect HWSV coil/heat exchanger leaking (as above).	Replace HWSV.
	Water levels too high in C/H feed and expansion cistern allowing hot expanded water to reach overflow outlet once system is heated.	Adjust ball valve to provide correct water level.
	C/H feed and expansion cistern too small to accept volume of expansion water when system is heated.	Exchange feed and expansion cistern for correct size.

Table 8.1 Fault finding chart – Systems (continued)

Symptom	Possible cause	Remedy
System noisy.	'Air' in system.	Implement recommendations as outlined under **Corrosion and noise – two common problems** in this Part.
	Trapped air in circulating pump.	Release trapped air from pump (see pump manufacturers instructions).
	Pump reverberating on wooden floor boarding or plasterboard walls.	Replace pump support brackets with ones containing rubber insulating bushes.
Replacement pump fails to circulate water.	Air trapped in pump.	Remove air by following manufacturer's commissioning instructions.
No air or water flowing from air release valves, on a system incorporating a close-coupled feed and vent arrangement. NB. Feed and expansion cistern may contain water.	Cold feed blocked at junction with system pipework. Suspect dissolved oxygen entering system.	Increase cold feed pipe size from CH feed and expansion cistern so that it contains at least 3% of the system water (see **Corrosion and noise – two common problems** in this Part).
Radiators not very hot.	Boiler thermostat defective.	Replace thermostat.
	Boiler manifold injector tee not fitted (where required) or defective on gravity domestic boiler installations.	Replace or fit new injector tee in accordance with the manufacturer's instructions.
Single pipe systems: Radiator(s) not fully heating up.	Boiler thermostat defective.	Replace thermostat.
	Bore of single pipe circuit too small.	Increase pipe size.
	Single pipe circuit too far from radiator(s).	Replace single pipe circuit pipework as close as possible to radiator connections.
	Radiator valves closed or partially closed.	Both radiator valves should be fully open.
	Incorrect thermostatic valves used i.e. two pipe system valves used.	Replace valves with those suitable for a single pipe system. This may involve extending the pipework to a top radiator connection and using inline single pipe thermostatic valves.
	Radiator thermostats shut down after a short period of operation (room still cold). Incorrect valves (see above, also bottom radiator connection used) thermostat sensing head affected by hot air currents from system pipework.	As above.

Table 8.1 Fault finding chart – Systems (continued)

Symptom	Possible cause	Remedy
Two pipe system: Radiators not fully heating up – downstairs in particular.	System not balanced e.g. too much circulating water delivered to upstairs radiators (15mm or equivalent pipework).	Adjust lock-shield valves on all radiators until there is a temperature differential of 11°C between the flow and return connections to each radiator.
	Unrestricted flow to domestic hot water circuit.	Adjust lock-shield valve on the domestic hot water circuit to give 11°C differential between flow and return pipes.
	Old inline motorised control valve operational but defective – causing restriction to water flow. **Note: Setting the valve to its manual position and operating the system on the domestic hot water setting of the programmer can often prove this.**	Replace valve with modern type.
Radiator fitted with a TRV does not heat up.	During long periods without use (i.e. warm weather), a TRV operating spindle can seize in the closed position holding the valve against its seating.	Free the operating spindle and maintain the TRV as necessary.
Radiator does not heat up but flow and return pipework hot (twin entry valves).	Radiator valve internal flexible return pipe connection misplaced or missing.	Drain radiator and refit or replace missing pipe.
Radiators hot but with reduced convected heat.	Convection fins blocked with lint.	Clean fins using an appropriate boiler flue brush and vacuum cleaner.
Convector radiators hot but with reduced convected heat.	Convection panel fins blocked with paint and/or lint.	Clean as appropriate.
Skirting radiators hot but with little or no convected heat.	Carpet laid and not secured under radiator blocking off air circulation path.	Cut and secure offending carpet.
	Heat exchanger fins blocked with lint.	Clean fins.

Table 8.1 Fault finding chart – Systems (continued)

Symptom	Possible cause	Remedy
Clicking noise from under floors when system is heating up or cooling down.	Circulating pipework in contact with wooden joists and/or floors.	Insulate offending pipework from woodwork.
Electrical room thermostats slow to respond resulting in variations in room temperature.	Room thermostat incorrectly wired. NB. Room thermostat fitted with an accelerator/anticipator may require the connection of a neutral wire to operate successfully.	Rewire thermostat in accordance with the manufacturer's instructions. This may involve replacing cables to the thermostat with one containing a suitable neutral cable. Alternatively, exchange thermostat for a modern type utilising a thermistor. This type of thermostat may not require a neutral wire.
	Room thermostat installed in the same room where a radiator is also installed which is controlled by a radiator thermostat (TRV). NB. Such systems are installed to avoid the continuous running of the circulation pump during programmed heating 'ON' periods. The radiator in the room with the room thermostat should not be fitted with a TRV.	Remove or immobilise radiator thermostat in fully open position.

Table 8.2 Fault finding chart – Boilers

Symptom	Possible cause	Remedy
Boilers general:	Pilot light too small.	Clean and/or adjust pilot flame.
Pilot light goes out.	Defective thermocouple.	Replace.
	Room-sealed flue terminal inlet/outlet blocked or obstructed with undergrowth/spiders webs.	Clear obstruction.
	Room-sealed flue terminal in re-entrant position i.e. POC re-entering appliance air inlet duct.	Remove/re-site cause e.g. drainpipe etc. or re-site boiler and flue terminal (see Essential Gas Safety – Domestic – Part 13).
	System pump fails, causing overheat device to operate, breaking electrical connection to thermocouple.	Replace pump and, where appropriate, reset overheat device.
	Overheat device defective.	Replace overheat device, – on NO account must the device be bridged out i.e. by-passed.
	Overheat device operates – no system bypass or bypass incorrectly adjusted.	Fit or adjust bypass in accordance with the boiler manufacturer's instructions.
	Boiler internal flue seals defective allowing POC to mix with incoming fresh air (vitiated atmosphere).	Replace defective seals.
	Draught diverter missing on open-flue boiler. NB. This will result in excessive draught through the boiler.	Fit manufacturer's draught diverter.
Boiler fails to ignite.	No electricity.	Investigate cause.
	Boiler thermostat or room thermostat not operating/defective.	Adjust setting or replace as appropriate.
	Boiler fan not operating.	Investigate cause and replace as necessary.
	Multifunctional control valve solenoid not operating.	Investigate cause and replace valve or, where appropriate, solenoid operator.
	Motorised valve not operating to provide electricity supply to boiler.	Investigate cause and replace valve or, where appropriate, the synchronous motor.
	Pilot alight but vapour valve fails to open – pilot too short or valve defective.	Clean pilot injector – replace vapour valve as appropriate.

Table 8.2 Fault finding chart – Boilers (continued)

Symptom	Possible cause	Remedy
Combination boilers – basic faults: Boiler fails to ignite.	No electricity.	Programmer/room thermostat in 'OFF' period or fuse blown.
	Inadequate gas pressure or no supply.	Check gas is 'ON' at meter emergency control valve. Check operating pressure at meter and boiler inlet connection (see Essential Gas Safety – Domestic – Part 7 for further guidance).
	Air in gas supply.	Purge gas supply pipework (see Essential Gas Safety – Domestic – Part 15 for further guidance).
	System pressure too low.	Re-pressurise system in accordance with the boiler manufacturer's instructions.
	Frequent system pressure loss.	Investigate cause (suspect worn radiator valve glands). Check for correct air or nitrogen pressure in the sealed system expansion vessel.
	Inadequate inlet water pressure.	Investigate cause.
	Appliance water filter blocked.	Clean as necessary.

Corrosion and noise – two common problems

A common problem of wet central heating systems is noise from trapped air circulating within the system - and therefore the need to vent air from radiators in upper rooms

This 'air' is often actually hydrogen gas - created by a reaction between oxygen in the water and various metals within the system. A by-product of this corrosion process is the formation of ferrous hydroxide and hydrogen. The ferrous hydroxide is slowly converted to magnetite and more hydrogen

The magnetite forms a black sludge, which settles in the low points of the system and can block circulating pumps, radiators and pipework

Air can enter the system from a number of sources. The most common are:

- leaking radiator valves and fittings on a negative pressure system
- dissolved oxygen in the water from the feed and expansion cistern on a positive/negative pressure system
- water from the open vent pipe 'pumping over' into the feed and expansion cistern on a positive pressure system creating oxygen rich system water
- water from the open vent pipe 'pumping over' momentarily into the feed and expansion cistern, gradually creating oxygen rich system water

number - see over.

Figure 8.3 Importance of circulating pump position

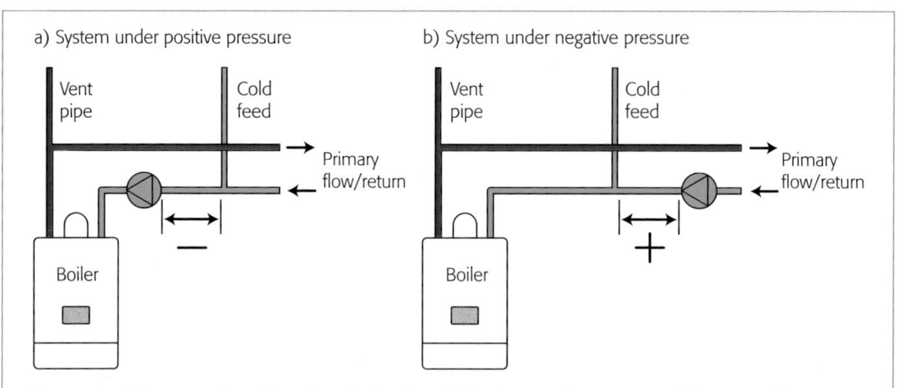

When a circulating pump is positioned at a), after the cold feed, a negative pressure will be created between the pump and cold feed whilst the rest of the system is likely to be subjected to positive pressure, with less likelihood of air entrainment.
With a circulating pump positioned at b), before the cold feed, a positive pressure will be created between the pump and cold feed, whilst the rest of the system is likely to be subjected to negative pressure, allowing air to leak into the system.

In the case of 1 – a negative pressure system (caused by incorrect positioning of the circulating pump). This can cause air entrainment into the system through fittings, which show little signs of water leaks but allow air to 'leak' back into the system once the circulating pump is 'running' (see Figure 8.3).

In the case of 2 – once the system is turned off, water in the feed and expansion cistern – which has absorbed oxygen from the atmosphere – cools down and contracts back into the system. It then enters the system circulating pipes via the cold feed pipe. If this movement of dissolved oxygen in the water can be contained within the feed and expansion cistern and cold feed pipe, it is unlikely to enter the circulating pipes and cause corrosion.

- you can achieve this by increasing the cold feed pipe size and/or length so it contains at least 3% of the system water content
- in the average 3 bedroom semi-detached household, this generally means increasing the cold feed pipe size from the 'traditional' minimum 15mm to 22mm diameter

In the case of 3 – water in the open vent pipe which has been subjected to positive pressure (caused by the position of the circulating pump in relation to the open vent and cold feed arrangement) 'pumps over' into the feed and expansion cistern, stirring up the sediment - and again introducing air into the water.

This creates a circuit of its own: causing this water to circulate around the system and continually pump over into the feed and expansion cistern.

- you can generally overcome this by correctly locating the circulating pump (see **System requirements** in this Part)

In the case of 4 – water in the open vent pipe can be subjected to positive pressure (movement). This can happen momentarily - when the circulating pump is turned on, or off. The sudden movement of the water causes water to be pushed out of the open vent and into the feed and expansion cistern.

- you can generally overcome this by raising the height of the open vent pipe above the feed and expansion cistern, to take up the general movement in the water level caused by the operation of the circulating pump - or you can incorporate a surge arrester in the open vent pipe (see **System requirements** in this Part)

Note: Once the circulating pump is 'running', radiators are subject to negative pressure and may admit air to the system if the radiator air release valve is opened - as opposed to releasing water.

Figure 8.3, depicts a two-pipe system showing two different circulating pump and cold feed positions.

With the circulating pump positioned, as in Figure 8.3(a), the system operates under a positive pressure - with less likelihood of air entrainment. But with a circulating pump positioned as in Figure 8.3(b) in front of the cold feed connection, with the circulating pump running, this creates a negative pressure - and air may be entrained into the system.

Reports of fumes from open-flued appliances – what you must do

When you investigate reports of fumes, or there are signs of spillage from a particular open-flue appliance, but there is no evidence of spillage when you test the appliance, under certain conditions (e.g. doors or windows to adjoining rooms open) the open-flued chimney serving the affected appliance may be subject to thermal inversion (see the current Essential Gas Safety – Domestic – Part 14 for further guidance)

In such cases, if the dwelling contains other heating appliances (e.g. a central heating boiler, gas fire or electric storage heaters etc.) and it is particularly air tight, carry out a spillage test with these appliances operating and the whole house heated.

- when you test an installed open-flued gas appliance in a dwelling that is unheated and relies on adventitious ventilation for the flue to operate correctly, under certain conditions it may pass a spillage test
- if you apply the same test when the dwelling is heated, the appliance may be subject to thermal inversion

This is more of a problem with appliances that are installed in buildings with greater air tightness. For example:

- timber frame buildings or particularly well draught-proofed dwellings; and/or
- where the appliance you are testing is flued into a chimney on an external wall - or has a chimney that is routed externally

Where an open-flued appliance is affected this way, if you provide additional ventilation to the room where the appliance is installed, it may solve the problem.

See the current Essential Gas Safety – Domestic – Part 14 and 'Using Electronic Combustion Gas Analysers for Investigating Reports of Fumes' for further guidance.

Reports of fumes from room-sealed appliances – what to do to avert danger

Natural-draught room-sealed appliances:

- make sure the outer case of a natural-draught room-sealed appliance is fitted correctly
- renew any sealing material as necessary
- ensure that the case itself fits securely and that all fixing bolts/screws are located correctly
- ensure chimney system is correctly assembled/joined (see Note)
- hazardous incidents can occur as a result of poorly assembled/maintained room-sealed appliances, where POC are able to escape into the room

Fanned-flue room-sealed boilers – hazardous instances have occurred where POC have leaked from positive pressure fanned-flue room-sealed boilers (see **Testing appliances with positive pressure cases – steps to help you ensure safety** below).

Note: This is a particular concern where chimney systems are routed through voids (see 'Chimneys in voids' in this Part).

Testing appliances with positive pressure cases – steps to help you ensure safety

Many gas appliances in customers' homes still use positive pressure case technology. But the gas industry recognises the danger from inadequately sealed positive pressure case appliances. So there is an agreed procedure to help you recognise the signs and assess the condition of this type of appliance.

What are 'Positive Pressure Gas Appliances'?

Historically, fanned draught room-sealed boilers were of the positive pressure type. For a positive pressure appliance to operate safely, the combustion chamber casing must be firmly secured to the boiler chassis, as the manufacturer intended with the correct seal in a good condition (see Figure 8.4)

What happens if this is not done?

- there is a real risk that POC may escape into the room in which the appliance is installed
- due to the poor combustion that is likely to occur, high levels of carbon monoxide (CO) could be produced - creating a dangerous environment

Figure 8.4 shows the differences between positive and negative pressure appliances.

Testing: what you must do

Regulation 26(9) of the Gas Safety (Installation and Use) Regulations (GSIUR) requires you to test the effectiveness of any flue after you have worked on a gas appliance.

The Industry test method to help ensure that case seals of positive pressure gas appliances comply with the requirements of the GSIUR is as follows:

Step 1

Before you put the case back on the appliance, check:

- are any water leaks evident?
- is the backplate or case corroded?
- where you see corrosion, is it likely to affect the integrity of the case, backplate, or seal?

Figure 8.4 Positive/negative fanned flued room-sealed gas appliances

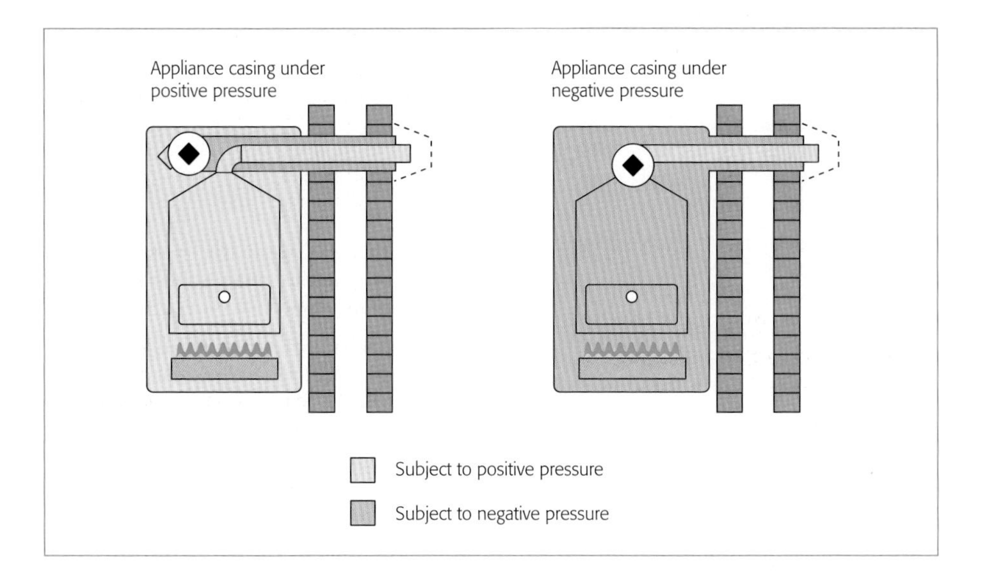

Note: Carefully check the extent of the corrosion with a sharp instrument e.g. a screwdriver. If the instrument doesn't perforate the corroded area, take this as acceptable. But advise the gas user of the problem - and potential consequences if the repair is not made.

- are the combustion chamber insulation linings intact?
- is the backplate or the case distorted or damaged? Pay particular attention to the area where the case and seal meet. Explosive ignition of the main burner may have caused it.
- is the case sealing material intact and in good condition? (e.g. pliable, free from discolouration, trapped debris, etc.). Will it continue to form an adequate seal between the case and the backplate?
- is anything trapped or likely to be trapped when you put the case back on (e.g. wires, thermocouple capillaries, tubes, etc.)?
- are other gaskets and seals intact?
- is the pilot inspection glass undamaged?
- are the case fastenings and fixings (including fixing lugs) in good condition? (e.g. screws/nuts stripped)
- are there any signs of discolouration on or around the appliance, which may have been caused by leaks of POC from the appliance?

Rectify any defects you identify in Step 1 as necessary and proceed to Step 2.

Note: If you identify defects, use the current Gas Industry Unsafe Situations Procedure to classify them:

- **if case fittings are inappropriate or missing or seals defective – and you cannot remedy this, but there is no evidence of leakage - classify the appliance as At Risk (AR)**
- **if there is evidence of actual leakage, then classify the appliance as Immediately Dangerous (ID)**
- **if suitable replacement seals are no longer available, class the appliance as ID and regard it as obsolete**

Step 2

When you have put the case on the appliance, check:

- is the case fitted correctly?
- is a 'mark' visible, showing that the case was previously fitted closer to the backplate?
- are all the case screws adequately tightened?
- is a bright area visible on the screw thread of any of the case securing screws, indicating that the screw was secured more tightly before?
- is anything trapped and showing through the case seal?

Rectify any defects identified in Step 2 as necessary. Proceed to Step 3.

Step 3 – Operate/light the appliance

- ensure that the main burner remains lit (i.e. set the appliance and room thermostats to their highest settings)
- check for possible leakage; do this initially by running your hands around the boiler casing and backplate
- then check for possible leakage etc. as in Step 4 where practicable

Step 4 – Check for possible leakage of POC from the appliance

- if joints have been disturbed, use LDF to confirm there are no gas escapes
- check for possible leakage of POC from the appliance by using a taper (to get into less accessible locations), an ordinary match, or similar

Note: Although you can use smoke tubes and smoke matches, the results may need further interpretation.

- light the taper/match and allow the flame to establish. Put the flame very close to the case seal or any possible leakage point (e.g. back panel)
- any draught caused by a leak will blow the flame quite easily. Move the taper around the entire seal (use fresh tapers if needed)
- to check the seal at the bottom of the case – hold the lit taper between the bottom of the case and the appliance control panel. Does the flame flicker slowly - or is it disturbed by leakage flowing from the case? Try the taper in several positions

Attention: DO NOT confuse natural convection with leakage. DO NOT look for a gas escape with this method.

- rectify any defects as necessary and re-check
- if still unsure seek expert advice

Note: You must be careful not to set fire to surrounding fixtures/furnishings.

Update to the procedure

The Health and Safety Executive (HSE) advise that you can use the smoke producing device/wand that produces a steady thin stream of smoke now available, instead of a lighted taper/match - and this is now the preferred method.

But if you have any doubt about the suitability of any smoke producing device/product (particularly with reference to COSHH requirements) follow the original outlined procedure.

How to classify the installation

BS 5440-1 has been reviewed and the issue about testing positive pressure case seals is included as an Appendix. There is additional text in the form of a 'Commentary on 10.4.2' which states:

"The appliance certification criteria permit a limited amount of case and seal leakage due to manufacturing tolerances. It is therefore likely that some minor leakage might be identified on positive pressure fanned flue boilers, in particular where a thermocouple lead/thermostat capillary or ignition high tension lead etc. passes through a grommet/gasket, or where there is a metal fold/joint that forms a corner on the boiler case itself."

"In these instances it is necessary to assess whether the leakage is due to normal manufacturing tolerances or to a defect with the grommet/gasket taking into account any previous customer reports of fumes, signs of staining, condition of the grommet/gasket etc, before deciding that the leakage identified is due to normal manufacturing tolerances and whether the appliance is safe to leave in operation."

"If there is any concern as to whether the level of leakage is significant and providing the point of leakage is not due to a defect in the main boiler case seal, e.g. around a grommet or gasket, it might be possible to effect a permanent repair by supplementing the original grommet/gasket with high temperature silicone sealant."

"It is essential that perforations in the case material due to corrosion are not temporarily repaired and any defective main boiler case seals requiring replacement should only be replaced with the manufacturer's supplied or authorised component".

It is essential that where you determine that any leakage is greater than normal manufacturing tolerances and you cannot rectify this at the time:

1. you classify the appliance as 'Immediately Dangerous' (ID); and
2. make it safe in accordance with the current Gas Industry Unsafe Situations Procedure

Table 8.3 contains a list of room-sealed fan assisted positive pressure gas appliances.

This list is not exhaustive - but use it as guidance for appliances that are believed to operate under positive pressure.

Warning – 'Immediately Dangerous' (ID) installations: your responsibility

- it is essential that in ALL cases where you identify spillage of POC, you make the appliance safe
- in the case of a new installation, if you cannot rectify the fault, you are responsible for disconnecting and sealing the appliance from the gas supply immediately - and labelling the appliance accordingly
- with existing installations, you must seek permission from the gas user to seal the supply. If permission is refused, turn off the appliance
- in both instances adhere to the current Gas Industry Unsafe Situations Procedure (see also the current Essential Gas Safety – Domestic – Parts 8 and 10)

Table 8.3 Room-sealed fan assisted positive pressure gas appliances

Manufacturer and model	Manufacturer and model
Alde International (UK) Ltd	**Halstead Heating & Engineering Ltd**
Alde 2927 Slimline	Halstead 45F*
Brassware Sales Ltd	Halstead 55F*
Ferrolli 76 FF*	Halstead 65F*
Ferrolli 77 FF*	Wickes 45F*
Crosslee (JLB) (Pyrocraft)	Wickes 65F*
AWB 23. 09 WT Combi	Barlo Balmoral 45F*
Crosslee (Trisave Boilers Ltd)	Barlo Balmoral 55F*
Trisave Turbo T45*	Barlo Balmoral 65F*
Trisave Turbo T60*	**Harvey Habridge Ltd**
Trisave Turbo 30*	Impala MK 11
Trisave Turbo 22*	Impala MK 11 Ridgeseal
Glow Worm Ltd	Impala Super 2 (HF)
Economy 30F*	Impala Super 2 (VF)
Economy 40F*	**Potterton Myson Ltd**
Economy 50F*	Myson (Thorn) Olympic 20/35F ‡
Glow Worm Fuelsaver 35F*	Myson (Thorn) Olympic 38/50F ‡
Glow Worm Fuelsaver 45F*	Myson (Thorn) Apollo Fanfare 15/30
Glow Worm Fuelsaver 55F*	Myson (Thorn) Apollo Fanfare 30/50
Glow Worm Fuelsaver 65F*	Supaheat 50/15 with 'A' control
Glow Worm Fuelsaver 80F*	Supaheat GC 50/15
Glow Worm Fuelsaver 100F*	Netaheat MK 1 10/16
Glynwed Domestic & Heating Appliances Ltd	Netaheat MK 1 16/22 BF
AGA A50	Netaheat MK 11 10/16 BF
AGA A50 A	Netaheat MK 11 16/22 BF
AGA A50 NG	Netaheat MK 11F 10-16 BF
AGA A50 SS	Netaheat MK 11F 16-22 BF
AGA A50 ANG	Netaheat Electronic 6/10
AGA A60	Netaheat Electronic 10/16
AGA A60 NG	Netaheat Electronic 16/22
AGA A75 NG	Netaheat Electronic 10/16e
Hi-light P50	Netaheat Electronic 16/22e
Hi-light P50A SC	Netaheat Profile 30e
Hi-light P50S	Netaheat Profile 40e
Hi-light P50SS	Netaheat Profile 50e
Hi-light P50S/A	Netaheat Profile 60e
Hi-light P50S/A GLC	Netaheat Profile 80e
Hi-light P50S/A SC	Netaheat Profile 100e
Hi-light P50/A	
Hi-light P70	
Hi-light P70S	
Hi-light P70SS	

Table 8.3 Room-sealed fan assisted positive pressure gas appliances (continued)

Manufacturer and model	Manufacturer and model
Stelrad Group Ltd Ideal Elan 2 40F* Ideal Elan 2 50F* Ideal Elan 2 60F* Ideal Elan 2 80F* Ideal Excel 30F* Ideal Excel 40F* Ideal Excel 50F* Ideal Excel 60F* Ideal Sprint 80F* Ideal W2000 30F* Ideal W2000 40F* Ideal W2000 50F* Ideal W2000 60F*	**Worcester Bosch** Heatslave 9.24 RSF* Worcester 9.24 Electronic RSF* Worcester 9.24 Electronic RSF 'S'*

‡ A safety enhancement kit has been designed for these appliances and is available from Potterton Myson Ltd.

* Boilers where spares relevant to case seal problems are still available, based on information provided by manufacturers.

Flame reversal – may be dangerous

Flame reversal is a condition where burner flames are distorted from the face of the burner. In the case of a gas fire connected to a back boiler/back circulator, the distorted flame burns under a gas fire so that the user may not easily see them (see Figure 8.5). It is often brought to your attention after a complaint about poor radiant effect from the gas fire.

- the condition can be dangerous - and is generally a result of excess air passing into the builder's opening or chimney
- this excess air can be as a result of a tall chimney creating too much up-draught (flue pull)
- more often in the case of a gas fire/back boiler installation, it is the result of an inadequate bottom seal between the flexible metallic flue liner and brick/masonry chimney.

  Or it is due to other openings within the builder's opening e.g. a ventilation opening, under floor ventilation or pipe ducts etc.
- another cause may be a fender that is too high or is fitted too close to the gas fire, thus preventing air from passing under the fire

The condition can be particularly aggravated on windy days.

Note: Due to the method of installation (i.e. a closure plate), a gas fire/back circulator is seldom subject to flame reversal. But if there is a complaint about poor radiant effect from the gas fire, do investigate the condition.

Under normal operating conditions for a gas fire/back boiler, air passes into the builder's opening and chimney under the fire and through the fire heat exchanger and air relief openings on each side of the fire.

Figure 8.5 Flame reversal (gas fire)

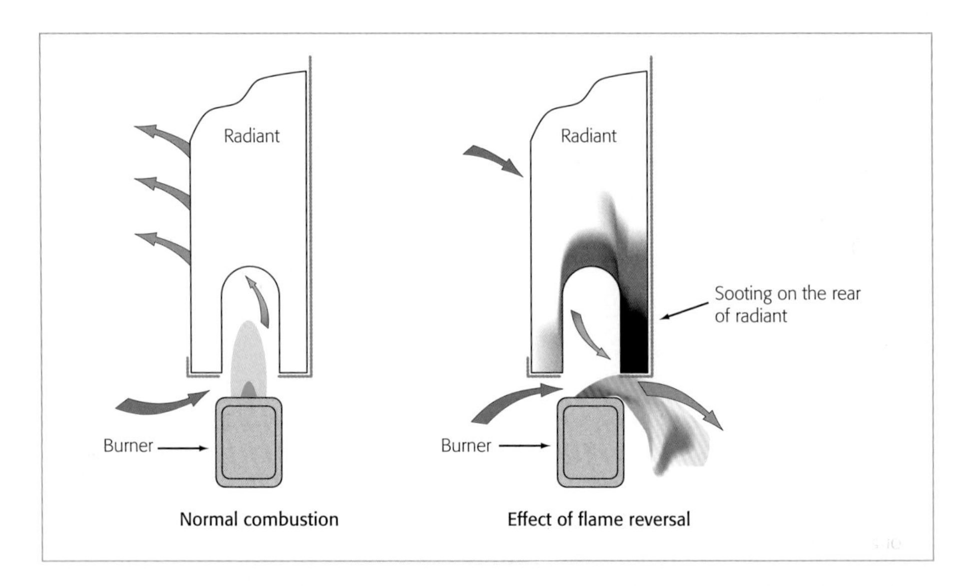

In abnormal conditions (e.g. tall chimney or adverse weather conditions creating an excessive up-draught in the chimney, or other openings within the builder's opening), the excess air will enter the builder's opening from the room by the most direct route, which is generally under the fire and/or downwards through the radiants.

When air passes downwards through the radiants it goes through the gap between the radiant support plate and burner, before finally passing to atmosphere or other rooms. If the gas fire is 'on' at this time, the burner flames will be distorted and will follow the same route as the air, burning under the fire – hence the term: flame reversal.

- when you investigate a complaint about poor radiant effect from the gas fire and you suspect flame reversal, it is particularly important that the fire front case is in position and completely assembled i.e. boiler control access panel in position where applicable
- if you fail to place this panel in position, it will give misleading results (e.g. the excess air will be diverted through the opening and not through the radiants – so flame reversal may not occur)
- also, during your visit the wind conditions may not be right to create the effect. However, a gas fire subject to flame reversal may bear some visual evidence of the condition

The following is a list of possible signs of flame reversal:

1. The fire front chrome fender trim may be 'blued' or distorted, or
2. The burner may show signs of distress e.g. it may be discoloured and have surface cracking or severe metal fatigue and corrosion, or
3. The air guide or radiant support plate may be distorted and might show signs similar to 2 above, or
4. The ignition lead may be burnt, or
5. There may be severe distortion and/or splitting of the heat exchanger, or
6. The front bottom outer case may be burnt or distorted.

If you are instructed to repair the fire, you must thoroughly examine the appliance.

Unless you are absolutely certain that only superficial damage has occurred, you must treat it as 'ID' in accordance with the current Gas Industry Unsafe Situations Procedure (see also the current Essential Gas Safety – Domestic – Parts 8 and 10).

A damaged burner, a misaligned air guide or radiant support plate, or a severely damaged heat exchanger etc. may lead to flame impingement and the production of CO which may escape into the room from the damaged appliance.

Wall staining – always check cause

Discolouration of wall surfaces:

Above the fire

All space heaters generate warm air convection currents and transfer heat to any wall surface against which they are placed.

Some modern vinyl wall coverings are affected by heat and are easily discoloured by that heat, which can be confused as staining caused by the spillage of POC into the room.

- when you investigate such a complaint, be aware of this possibility
- on no account dismiss the report of fumes, as you must always carry out a thorough investigation following all reports of their possible presence

Upper rooms

Some wall staining is also caused by acid attack from within the brick/masonry chimney, caused by water (rain or condensation) mixing with soot to form an acid. This may penetrate the brickwork and appear as a dark brown stain on upper room/bedroom walls, spoiling the décor.

- in this case, if a flexible metallic liner is installed this will help keep the POC warm and above 'dew point' (a point at which condensate forms) - thus eliminating the problem

Odd coloured gas flames

Some gas appliances covered by this Part have burner flames that are visible: for example, open-flued boilers and gas fire/back boilers/circulators.

In the case of gas fires, the burner flames either heat radiants or create a realistic effect (by passing through a fuel bed simulating solid fuel). In most cases the burner flames are exposed to the air in the dwelling and this does not generally present a problem.

However, if a householder suffers with a respiratory condition and uses a nebuliser to relieve the symptoms, be aware that the gases given off by the respiratory device may affect the flame characteristics and hue. Depending on the concentration of these gases in the room, they may cause the burner flames to change colour. The colour range may vary from a pale pink to a bright orange.

The flames can also appear to be much larger than normal, as salts in the gases expose the full outer mantle of the flame, which is normally not visible to the naked eye.

The flames will return to their normal characteristic size and colour once the room(s) affected by the gas from the nebuliser have been purged with fresh air.

Electrical connections – you are responsible for compliance

- all electrical work must comply with the Electricity at Work Regulations
- electrical connections to the appliance must conform to BS 7671: Requirements for Electrical Installations. IET Wiring Regulations Seventeenth Edition
- you must ensure all electrical work is carried out by a competent person: for example, operatives approved by the National Inspection Council for Electrical Installation Contractors (NICEIC)
- where you route installation wiring in an appliance compartment, do note the higher ambient temperatures which may exist in these situations
- make sure that electrical wiring is not subject to temperatures in excess of its specified rating

Electrical isolation – essential for safety

- provide electrical isolation so that all voltage can be effectively 'cut off' to prevent or remove danger whilst undertaking any work on the appliance
- it must also provide an efficient, easily operated means of disconnection – so site it to prevent danger
- install the electrical supply point in a readily accessible position, as close as practicable to the appliance (usually 1.5m). Connect it in accordance with the manufacturer's installation instructions with regard to:
  - correct method and polarity
  - fuse rating
  - earth connection and voltage range

The method of connection needs to provide electrical isolation by either:

1. A fused, double-pole switch or connection unit, or
2. A fused three pin plug and a shuttered socket outlet (except in rooms containing a fixed bath or shower – see **Gas appliances in bathrooms – electrical zoning requirements** in this Part).

Note: In the case of 2, remove the electrical plug when you service the appliance. To encourage this, an unswitched socket outlet is recommended.

Whichever means of electrical isolation you use, it must allow safe working on the appliance.

Your responsibility

- before you begin work on an appliance or installation, make sure that the installation is completely isolated from the electrical supply. To ensure this, use an approved test method (see **Safe isolation procedure** in this Part)
- also ensure that the supply cannot be restored without your knowledge, by doing one of the following:
  1. Where a fused double-pole switch or connection unit is fitted, withdraw the fuse carrier, remove the fuse and fit a small padlock to the carrier, in the open position.
  2. Where a plug and socket outlet is fitted, remove the plug from the socket.

These methods are suitable where the isolation is 'local' and within your sight.

- when the isolation is 'remote' attach a suitable warning notice to the means of isolation, stating: 'Danger Do Not Switch On'
- as a precaution, where a plug and socket is used, it is recommended you remove the fuse from the plug for safety or employ proprietary locking mechanisms such as plug lock boxes, which encase the plug entirely in a lockable plastic box

Safe isolation procedures

1) Remove the load from the circuit to be isolated.
2) Open the means of Isolation and secure the isolation device in the open position with a lock or other suitable means.
3) Remove and retain securely the circuit protective device (fuse) wherever possible.
4) Prove the correct operation of a suitable voltage indicating device on a known voltage source (proving unit or known supply).

5) Using the voltage indicating device check that there are no dangerous voltages present on any circuit, conductor or equipment to be worked on.

Test Between –

- Earth and line (L)
- Neutral (N) and line (L)
- Earth and neutral (N)

6) Re-prove the voltage indicating device as in 4 above.
7) Where necessary fit appropriate warning notices at the point of isolation.

Protective equipotential bonding – what you must do

- anyone who connects any installation pipework to a primary meter must, if protective equipotential bonding (PEB) proves necessary, inform the responsible person that such bonding must be carried out by a competent person
- the Gas Safety (Installation and Use) Regulations oblige you, when you install a section of pipework which connects the primary meter or emergency control valve, to inform the responsible person of the possible need for PEB
- You must do this:
  - even if this requirement did not exist before the work was undertaken
  - whether or not the meter and control are fitted
- electrical bonding must be carried out by a competent person
- give the advice in writing - CORGI*direct* produce a suitable label for use in this regard (see the current Essential Gas Safety – Domestic – Part 10 for further guidance)

Although the regulation applies only when you install new systems and modify existing ones:

- take similar action if you notice an apparent defect in electrical bonding in other circumstances, e.g. during maintenance checks (this applies to both protective or supplementary equipotential bonding)

You can see the positioning of PEB, where fitted to internal and external gas meter installations, in the current Essential Gas Safety – Domestic – Part 5.

Fireguards – what you may need to do

Apart from Decorative Fuel Effect (DFE) gas appliances and some Inset Live Fuel Effect (ILFE) gas fires, most gas fires are fitted with an integral dress guard. This guard is to prevent accidental contact between flammable material and the incandescent parts of the appliance. The dress guard may also help to retain artificial logs or coals from tumbling onto the hearth.

- where the gas fire is to be used in the presence of young children, the elderly or infirm you may need to screen it by adding a fireguard (CONSUMER PROTECTION – The Heating Appliance (Fireguards) (Safety) Regulations) and preferably attach it to the surrounding wall, fireplace surround etc.

Floor and wall protection

- if you install an open-flued gas boiler, any requirement to protect the floor or wall on which you mount the boiler will be detailed in the boiler manufacturer's installation instructions
- in the absence of instructions, provide a non-combustible insulating base of at least 12mm thickness under the boiler if the floor supporting the boiler is of combustible material

- in the case of a floor-standing boiler, also take care to ensure that the floor on which you place the boiler can support its weight
- also, should the floor be exposed to a prolonged period of wetness, due to a water leak for example, ensure that its strength would not be impaired (e.g. chipboard flooring supporting a boiler under these circumstances may collapse under its weight)
- equally, in the case of a wall-mounted boiler, the wall must be non-combustible - or the boiler must be suitable for installation to a combustible wall
- the wall must also be able to support the weight of the boiler - and use the correct number, size and length of fixing screws or bolts as recommended by the manufacturer

Note: When you install gas boilers in a timber framed dwelling, there are special requirements for this type of installation (see the current Gas Installer Manual Series – Domestic – Gas Installations in Timber/Light Steel Frame Buildings)

Chimney systems

Flue box/enclosure

If there is no suitable existing brick/masonry chimney, you can erect a false chimney breast - or a catchment space adapted to accommodate a suitable flue box or enclosure to which you can connect a combined gas fire/back boiler/back circulator (see Figure 8.6).

Both metallic flue boxes which comply with BS 715 and propriety flue block concrete enclosures are suitable for use with the back boiler/back circulator.

- using suitable materials, you may also construct on site an enclosure that meets the requirements
- always comply with manufacturer's clearance requirements
- only use rigid metallic chimney systems when you connect to a flue box within a false chimney breast – not a flexible metallic liner. A flexible metallic chimney liner is for use only within a brick/masonry chimney
- design the chimney in accordance with the chimney and flue box manufacturer's instructions (see also the current Essential Gas Safety – Domestic – Part 13 for further guidance)
- unless the manufacturer's instructions specify otherwise, install the fire/back boiler/back circulator to a minimum 125mm diameter flue
- you must stand the flue box and gas fire on a non-combustible base (see **Builders opening – Hearth** in this Part) and install and secure it in accordance with the manufacturer's installation instructions
- seal all openings, including gaps and cracks within the chimney breast, to prevent the escape of POC to upper rooms e.g. pipe or chimney openings in the ceiling etc.
- in the case of a back circulator, seal the closure plate to the front face of the flue box and install the fire and back circulator in accordance with the manufacturer's instructions (see also the appropriate guidelines of **Part 7 – Combined gas fires/back circulators** in this manual)

Note: Only use flue boxes identified as suitable for use by the back boiler/back circulator manufacturer and/or flue box manufacturer.

Figure 8.6 Method of installing a combined back boiler/back circulator, in conjunction with a flue box/enclosure and metallic chimney system

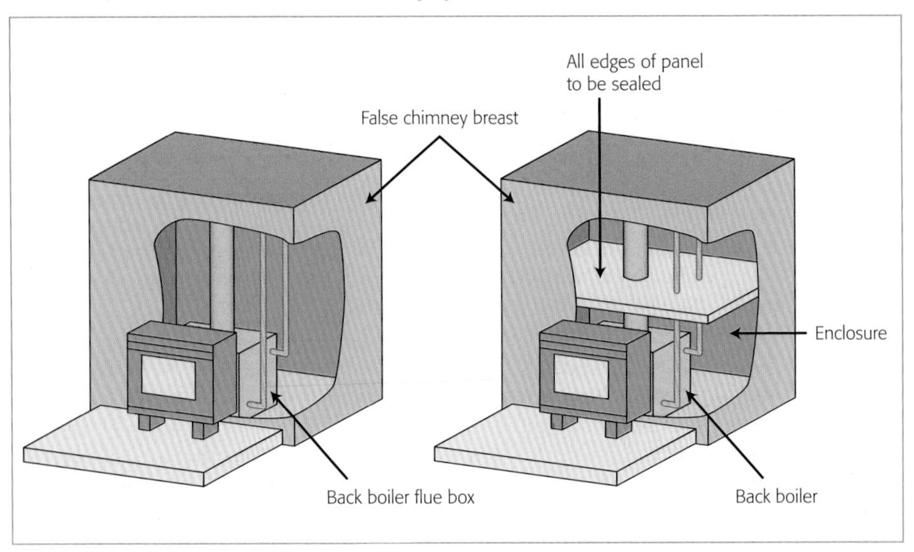

Chimney liner – its function

The liner prevents condensate (water vapour present in the POC which when cooled reverts to a liquid) from cooling and forming on the internal chimney walls. Once formed, the condensate will mix with the existing soot to form an acid. This may penetrate the brickwork and appear as a dark brown stain on upper room/bedroom walls spoiling the décor (see **Wall staining** in this Part).

In addition, if the POC are reduced in temperature, less heat is available and this could adversely affect flue performance (see the current Essential Gas Safety – Domestic – Part 14 for further guidance).

Flexible metallic flue liner

- when you install a flexible metallic flue liner, you must install the liner in one continuous length (no joints are allowed). The liner must be the same diameter as the boiler flue spigot outlet and have a minimum diameter of 125mm where it is serving a gas fire/back boiler/back-circulator
- connect the liner to the appliance flue spigot and secure using mechanical fixings (e.g. with a clamp and/or self-tapping screws); and
- seal the gap between liner and boiler flue spigot with fire cement, or by the method approved by the boiler manufacturer
- Methods of securing flexible flue liners – to the appliance and within chimneys – will typically be found in the liner manufacturer instructions

As a general guide:

- secure the liner at the top of the chimney, using the liner manufacturer's flue plate and clamp to support the weight of the liner. The clamp and plate should rest on, and be sealed to, the top of the chimney, leaving sufficient liner to rebuild the flaunching (cement/mortar slope) and fitment of an approved terminal
- the flaunching should make a weatherproof joint, leading water away from the liner and terminal. The terminal should always be fitted - and provide adequate space around it to ensure rapid dispersal of the POC

Note: Due to the age and condition of the brick/masonry chimney the chimney top may have deteriorated to the point where it is difficult for you to achieve a satisfactory seal at the top of the chimney.

In this case, it may be necessary to rebuild the chimney or replace the missing brickwork, which may involve the use of other trades.

How to seal the annular space between the flexible metallic liner and the brick/masonry chimney

- where the liner enters the builder's opening, seal the annular space between the liner and chimney (see image 'a' in Figure 8.7)

This seal is often referred to as a register plate or debris plate. The plate will certainly 'catch' any debris that falls down the chimney – but its main function is to prevent any movement of air passing up and around the liner to atmosphere.

In the case of a back boiler, air passing into this annular space should first pass under and through the fire into the builder's opening before passing up the flue to atmosphere.

On windy days, this movement of air may be excessive – in which case it will take a direct line, generally under the fire and/or down through the gas fire radiants to the builder's opening and flue.

If there is no seal, or an inadequate seal in place, this movement of air:

- will cool the room; and
- may disturb the flames on the gas fire causing flame reversal (see **Flame reversal** in this Part)

In the case of a back circulator unit, this may not apply because of the closure plate fitted to the front of the builder's opening.

- you need to make a seal between the liner and chimney. For a fire/back circulator installation, there is not generally a direct connection between the appliance and the flue opening (see image 'b' in Figure 8.7)
- in both cases it's common practice to make the seal using a non-combustible plate that you secure to the brickwork. Or if the annular space is not too large, you can seal using mineral wool tightly packed into the space
- if you do this, ensure the mineral wool will remain in position for the life of the installation. It's recommended you secure it in position - using a plate and clamp similar to the one used at the chimney top (see Figure 8.8)
- where the liner connects to the boiler flue spigot, secure and seal it in position in accordance with the manufacturer's instructions
- as a general guide, secure the liner using self-tapping screws and seal with fire cement

Figure 8.7 Method of using a flexible metallic flue liner in a brick/masonry chimney

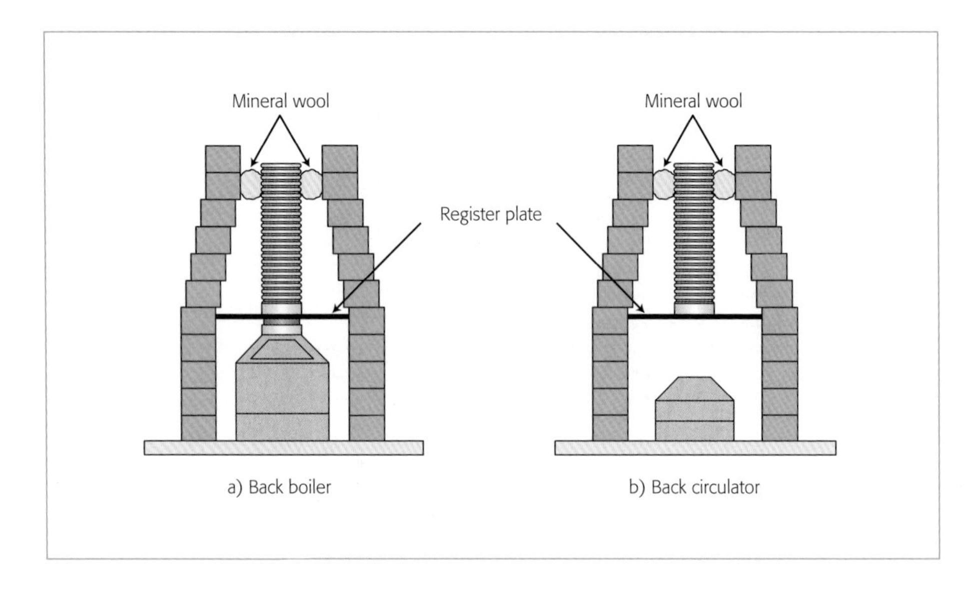

Figure 8.8 Method of sealing the annular space using mineral wool with plate and clamp

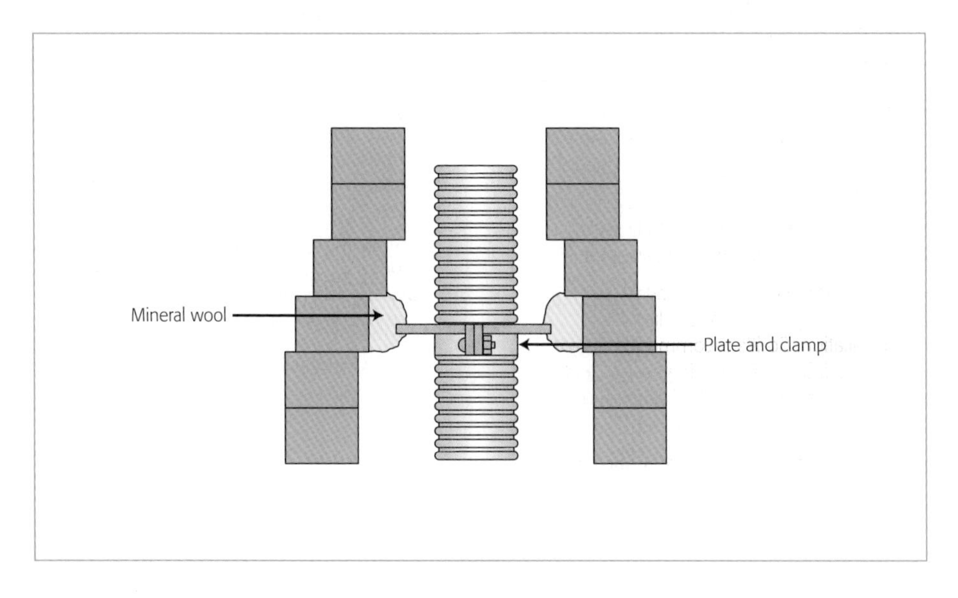

Figure 8.9 Method of securing and sealing a flexible metallic flue liner to a clay/ceramic liner

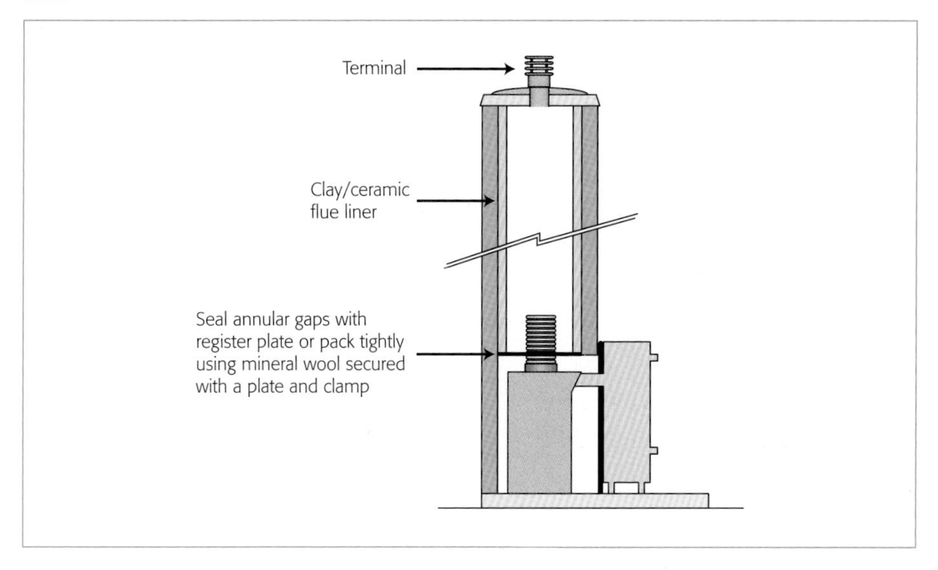

- a brick/masonry chimney built after 1965 is likely to be constructed with an earthenware liner. If so, you may not need to line the chimney throughout its length. Instead, in the case of a back boiler installation, simply take a short length of liner from the boiler flue spigot and pass it up into the earthenware liner a distance of 150mm (see Figure 8.9)
- it is important you seal the gap between the flexible metallic flue liner and salt glazed liner. You can do this by tightly packing mineral wool into the annular space
- if so, make sure the mineral wool will remain in position for the life of the installation. So it's recommended you secure it in position using a plate and clamp (see Figure 8.8) similar to the one used at the chimney top
- fit a flue terminal to the chimney top (see Figure 8.9)

Poured/pumped concrete chimney liners

These are acceptable alternatives to flexible metallic flue liners, but must only be carried out by a competent contractor.

- install poured/pumped concrete linings using a method certificated by an accredited test house
- always check that the lining is mechanically sound before you install an appliance
- sweep brick/masonry chimneys with this type of lining that have been used with another fuel - and carefully examine before use with a gas appliance

Terminals and noise – factors to consider

- which terminal you use impacts greatly on the inhabitants of the room in which the boiler is installed

For example, back boilers are generally installed in a living room - with a flexible metallic flue liner connecting the boiler to an approved terminal (generally aluminium). In most cases this combination is satisfactory.

- but if you install a flue terminal in a particularly exposed position, the terminal fins or holes can resonate in the wind, creating noise (this noise is often transmitted via the liner to the living area – and is amplified)
- most users can tolerate the noise as it is often only intermittent. But if the noise is intolerable, reduce it (and sometimes eliminate it) by fitting a terracotta terminal complying with BS EN 13502: 2002 'Chimneys. Requirements and test methods for clay/ceramic flue terminals'.

  These terminals are also particularly effective at reducing external noise

Test the flue for spillage of POC when installation complete

- when installation is complete, test the appliance according to the manufacturer's instructions to ensure that POC are not spilling into the room (see also the current Essential Gas Safety – Domestic – Part 14 for further guidance)
- generally, gas boiler and fire manufacturers have spillage testing instructions specific to a particular model. If so, you will find this information in their installation instructions – and/or on a data plate attached to the appliance

Flue testing with decorative re-circulatory ceiling fans present

- when you carry out a spillage test on open-flued appliances (particularly in the case of combined gas fire/back boiler/back circulators with an inset live fuel effect gas fire), if there is a decorative re-circulatory ceiling fan fitted in the same room, carry out spillage tests:
  - with the fan both on and off, at all speeds; and
  - where appropriate, in both directions

Tests show that these fans can disturb air movement in the room so as to cause spillage of POC - where none was present with the fan in the 'off' position.

Natural draught room-sealed boilers – terminal position critical

- choosing the terminal position on the outside wall is probably the most critical part of the installation
- for burner flames to burn correctly (complete combustion), POC must readily disperse and pass freely away from the concentric chimney terminal (combined air inlet and flue outlet duct) into the atmosphere - and not mix with clean fresh air passing through the air inlet duct to the burner
- when you install a boiler to harmonise with kitchen units, you must also consider the terminal position. It must not be restricted by an adjacent projection outside - e.g. buttresses, gate posts, soil pipes, internal or external corners of buildings etc.
- these are known as 're-entrant' positions i.e. the POC are prevented from being blown away by the wind, but re-circulate around the chimney terminal and may 're-enter' the appliance via the fresh air inlet duct. The POC vitiate this fresh air and reduce its oxygen content

- consequently, this has a profound effect on combustion quality at the burners and is likely (depending on the degree of vitiation) to cause the flames, including the pilot flame, to become ragged and lift off the burner
- in the case of the pilot burner, this could eventually lead to cooling of the thermocouple (permanent pilot models), which in turn could cause the appliance to fail to safety

The condition is particularly aggravated on windy days (see the current Essential Gas Safety – Domestic – Part 13 for further guidance).

- in the examples given, the condition may be regarded as unsafe – so if you find it, follow the current Gas Industry Unsafe Situations Procedure (see also the current Essential Gas Safety – Domestic – Parts 8 and 10)

Similar conditions apply if the concentric flue and air inlet duct is cut too short for the wall thickness. The air inlet grilles are then likely to be obstructed or blocked by cement mortar – which restricts air entrainment and adversely affects combustion.

- other 're-entrant' positions you may find are openings into buildings - such as doors, windows and ventilators
- some appliance manufacturers specify the dimensions where flues must be sited away from openings into buildings. Always comply with these
- if there are no specific instructions, ask the appliance manufacturer for guidance
- you must position the terminal so that the POC can safely disperse at all times e.g. when the termination is into a car-port or other similar structure, there must be at least two open, unobstructed sides to that structure
- pay attention to the material used on the roof - and provide allowance for adequate clearances/protection of the roof
- do not site terminals into a passageway, pathway or over adjoining property where they can be a nuisance or cause injury (see the current Essential Gas Safety – Domestic – Part 13 for further guidance)

Some natural draught room-sealed boilers are also suitable for installing onto Se-duct or U-duct chimney systems.

Fanned draught room-sealed boilers

Installation requirements are generally the same as those for natural draught room-sealed boilers, although it is less critical where you site a terminal for a fanned draught boiler. The fan helps disperse the POC.

- although siting requirements are more relaxed, you must take the same precautions as for natural draught room-sealed chimney terminations e.g. when the termination is close to openings into buildings or is into a car-port or other similar structure
- also ensure the POC and any pluming are not blown onto an adjacent property, causing a nuisance (see the current Essential Gas Safety – Domestic – Part 13 for further guidance)

The boilers are generally designed to incorporate several flueing options with side, rear or vertical flue outlet positions available.

Some fanned draught room-sealed boilers are also suitable for installing onto Se-duct or U-duct chimney systems.

Figure 8.10 Vertex flue, a special vertical fanned draught room-sealed chimney system

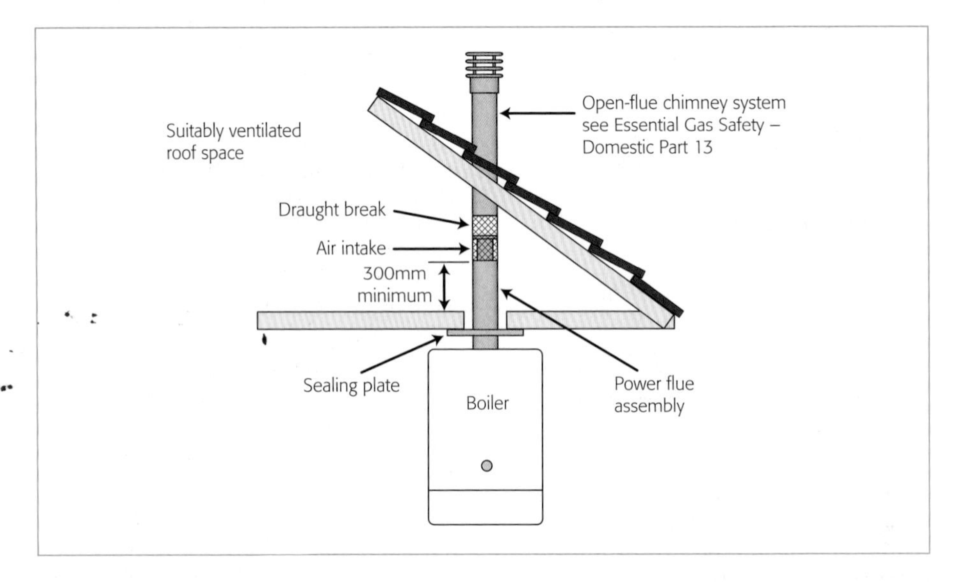

The 'Vertex' flue system

This is a variation on the vertical flueing option and is available for some types of boilers.

- with this design, you generally wall mount the boiler and position it close to the ceiling, adjoining the roof space
- a concentric chimney arrangement is taken directly from the boiler flue outlet to a position in the roof space, defined in the manufacturer's instructions
- this position must be at least 300mm above the level of any insulation material in the roof space
- at this position, fit a purpose-designed adapter (known as a 'draught break') to the concentric chimney system
- from the outlet of the draught break, install a conventional secondary chimney system - and terminate this through the roof, in accordance with the manufacturer's instruction
- protect the draught break with a guard in the roof space (see Note and Figure 8.10)

Note: This type of flueing arrangement takes its air for combustion from the roof space, which you must ventilate in accordance with the manufacturer's instructions (see the current Essential Gas Safety – Domestic – Part 4 for further guidance).

- where the roof insulation is of a loose fibrous material, make sure no airborne fibres are entrained into the appliance through the air inlet duct of the draught break

Se-duct and U-duct installations

When you plan to install appliances onto these types of chimney systems, check the two are compatible - and install in accordance with the manufacturer's instructions
(see also the current Essential Gas Safety – Domestic – Part 13 for further guidance).

Chimneys in voids

Modern gas boiler design has seen the introduction of new chimney materials (PVC for example), different chimney configurations (twin pipe as well as concentric) and consequently, longer chimney runs to the point of termination.

With the flexibility offered by modern chimney design, the choice of boiler position is no longer constrained to 'anywhere as long as it's an outside wall' and as such, boilers can be located almost anywhere in a property.

This fact is good news for developers/architects, particularly those of multi-occupancy dwellings (flats) where the boiler can be 'hidden' in a compartment/enclosure or other space not so frequently visited/viewed by the home owner.

The chimney system of these boilers is then routed within voids – ceilings, floor, wall, etc. – to the point of termination, either on an outside wall or, if vertical termination, the roof.

However, many of these chimney systems cannot be periodically inspected once in position and enclosed by finished décor, causing a real issue for gas operatives/businesses who have a duty of care to discharge under the Gas Safety (Installation and Use) Regulations 1998 (GSIUR).

Worse still, cases have been recorded whereby occupants of affected properties have been exposed to CO, including a fatality as a result of defective/broken chimney systems within these voids.

The incidents of CO poisoning has mobilised the industry – construction, gas, HSE, etc – to review current installation practices and to address the issues raised.

This review has culminated in new guidance being produced, both in the form of up-dated information included in to ADJ 2010 (1.47 Concealed flues) and manufacturers' installation instructions for new installations and via a Technical Bulletin (TB) issued by Gas Safe Register and a Safety Notice issued by the HSE for existing installations.

TB 008 (Edition 2.1) 'Room-sealed fanned-draught chimney/flue systems concealed within voids' provides important guidance which needs to be understood by gas operatives/businesses – visit https://engineers.gassaferegister.co.uk to download a copy of the TB.

Additionally, a supporting TB 008 (Edition 2.1) CIP-RACL has been produced to provide gas users and those responsible for the installation – landlords for example – guidance on the issues and what action is required to mitigate the risk posed.

New installations

ADJ requires that where chimneys are routed within voids that appropriate means of access be provided at strategic locations along the route of the chimney in order that periodic visual checks can be undertaken – servicing/maintenance visits for example.

The checks required are to confirm that:

- the chimney system (flue pipe and air duct, either concentric or twin pipe design) is continuous throughout its length
- all joints appear to be correctly assembled and sealed (refer to manufacturers instructions for guidance)
- the chimney system is adequately supported throughout its length

- the chimney system has the required fall back to the appliance (condensing appliances) and/or additional drain points, where required, are installed

The route of the chimney within the void needs to be confined to the dwelling concerned and not pass through other properties (communal areas are acceptable where inspection hatches are provided), given that access for inspection purposes cannot be guaranteed in these locations.

The means of access needs to be sufficiently sized to allow inspection, but they are not intended to be of such a size that would allow full physical access. An access hatch size of 300mm x 300mm is suggested, located at strategic points along the chimney route – changes of direction and within 1.5m of each joint (see Figure 8.11).

It's important that any access hatches do not impair the fire, thermal or acoustic characteristics required under Building Regulations. This may mean using access hatches with resilient seals and exhibiting the same characteristics for fire, heat and sound as the structure surrounding them.

Existing installations

TB 008 (Edition 2.1) is of real importance when dealing with existing gas installations, as it is highly likely that access provisions for inspection purposes hasn't been provided for.

The TB recognises this issue but importantly, doesn't want this situation to remain indefinitely and therefore, has set a cut-off date by which inspection hatches need to be installed or the gas installation concerned will be classified as 'At Risk' (AR) in accordance with the current Gas Industry Unsafe Situations Procedure and turned-off.

The cut-off date is 31st December 2012.

From now till then, gas operatives have a duty to inspect and test gas appliances as normal (Regulation 26(9) checks), but with an added dimension.

Regardless of whether inspection hatches have been provided or not, a thorough risk assessment of the installation needs to be undertaken. TB 008 (Edition 2.1) provides a checklist for the risk assessment process, which has been reproduced as a form, available through CORGI*direct* (see Figure 8.12) to assist gas operatives going about their work.

If after following and completing the risk assessment process an entry is made in to the non-shaded column (CORGI*direct* form) or the 'red' column on the TB, then the installation needs to be classified as 'AR' or 'Immediately Dangerous' (ID), as appropriate.

An entry in to the shaded column (CORGI*direct* form) or 'green' column on the TB means that the installation can be left in operation until inspection hatches are installed and before the cut-off date is reached.

Remember: Gas users/responsible persons have until 31st December 2012 to install suitable inspection hatches!

Additionally, and where inspection hatches are not currently installed, the risk assessment process also requires the installation of suitable CO alarms complying with BS EN 50291 in every room along the chimney route/suspected route – including any neighbouring property, as necessary.

Note: You cannot leave gas users/responsible persons to install suitable alarms at a later date. They must be in position at the time of your visit.

Ensure you obtain a copy of TB 008 (Edition 2.1), read and fully digest its contents.

Figure 8.11 Example locations of access panels for concealed horizontal chimneys in voids

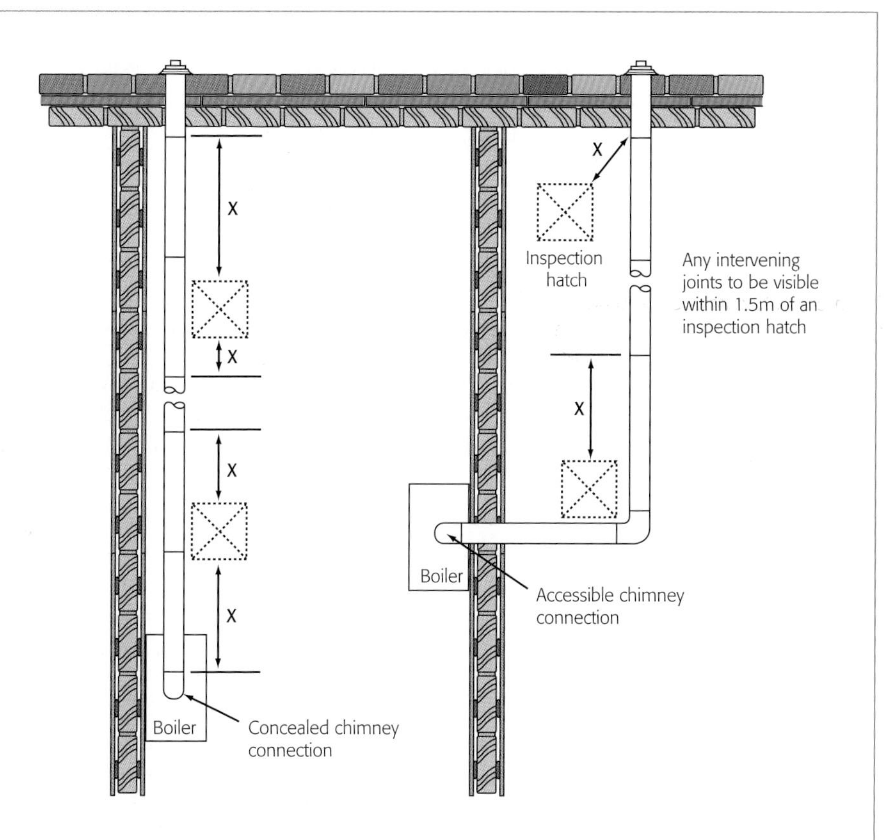

All voids containing concealed chimney systems should have at least one inspection hatch measuring at least 300mm square.

No chimney joint within the void should be more than 1.5m distant from the edge of the nearest inspection hatch, i.e. dimension ' X ' in the diagram should be less than 1.5m

Locate inspection hatch to optimise the ability to inspect the chimney system, e.g. inspection hatches located directly below the route of the chimney system may restrict inspection.

Where possible, inspection hatches should be located at changes of direction. Where this is not possible then bends should be viewable from both directions.

Figure 8.12 'Risk assessment for existing chimney systems in voids where inadequate access for inspection is provided' form

Customer Ref:

RISK ASSESSMENT FOR EXISTING CHIMNEY SYSTEMS IN VOIDS WHERE INADEQUATE ACCESS FOR INSPECTION IS PROVIDED

Gas Safe Register

This form should be completed in conjunction with Gas Safe Registers Technical Bulletin 008 (Edition 2.1): Room-sealed fanned-draught chimney/flue systems concealed within voids - visit www.gassaferegister.co.uk

Registered Business Details	Job Address
Registration Number:	Name (Mr/Mrs/Miss/Ms):
Company:	Address:
Address:	
Postcode: Tel No:	Postcode: Tel No.
Gas Operatives Name:	Rented Accommodation: Yes or No?
Operative Licence No:	Landlords Tel No:

If all answers are entered in the left hand (orange shaded) column of this Checklist, then the appliance may be left operational until means of access have been provided, or until **31st December 2012**. If any response is given to any question or statement in the right hand column of this Checklist, then the installation should be classified as 'Immediately Dangerous' (ID) or 'At Risk' (AR) as appropriate in accordance with the current Gas Industry Unsafe Situations Procedure (GIUSP).

REGARDING ACCESS FACILITIES TO CHIMNEY SYSTEM	YES	NO
1. Is it possible to determine the likely route of the whole chimney system?		
2. Where the chimney system is routed through neighbouring properties, is access available in the neighbouring properties to carry out this risk assessment or have reasonable steps been taken to ensure flue integrity?		
3. Is the ceiling or other enclosure free from evidence of distress or surface staining within the vicinity of the likely chimney route and which cannot be attributed to other causes (e.g. water leaks)?		
QUESTION FOR GAS USER/RESPONSIBLE PERSON	**NO**	**YES**
4. Is the responsible person/gas user and/or Gas Safe registered engineer aware of any previous history within the property, or other properties in the same development that could be related to chimney systems in voids issues that have not been corrected/rectified before completion of this risk assessment?		
REGARDING THE PRESENCE OF CARBON MONOXIDE (CO) ALARMS	**YES**	**NO**
5. Where Carbon Monoxide (CO) alarms are installed, can the gas user/responsible person confirm that there has been no history of alarm activation in the property?		
6. Are CO alarms conforming to BS EN 50291 installed/located in each room or internal space along the suspected route of the chimney system including where necessary neighbouring properties?		
7. Are the existing CO alarms installed in accordance with the manufacturer's instructions or industry guidance?		
8. Do the CO alarms 'alarm' when the test button is pressed?		
9. Will all CO alarms be within the manufacturer's recommended lifespan on **31st December 2012** or, where this information is not available, will they be less than 5 years old on **31st December 2012**?		
APPLIANCE OPERATIONAL CHECKS	**YES**	**NO**
10. Is the burner pressure and/or gas rate in accordance with the appliance manufacturer's specifications?		
11. Is satisfactory combustion performance being achieved? (See manufacturer's instructions, Gas Safe TB 126 and BS 7967-3 & 4). Where combustion performance analysis is not feasible but it is possible to inspect the flame picture, is the flame picture visually satisfactory?		
Record electronic combustion gas analyser readings (HR - high range & LR – low range), where appropriate: CO HR: ppm — CO_2 HR: % — CO/CO_2 Ratio HR: CO LR: ppm — CO_2 LR: % — CO/CO_2 Ratio LR:	–	–
12. Does the appliance appear to be functioning otherwise safely, e.g. all safety controls operating correctly, no signs of distress or staining around the appliance?		

RESULTS OF THE RISK ASSESSMENT (enter YES in the appropriate column)		YES		YES		YES
Using the above risk assessment in conjunction with the current GIUSP & TB 008 (Ed 2.1), the installation is:	Immediately Dangerous		At Risk		Left operational	

Gas operative signature: ______ Gas user/responsible person's signature: ______ Date: ______

Key: **Top Copy** – Gas User **Green Copy** – Gas Operative CORGI*direct* December 2011 To re-order quote Ref. CP43

To confirm the validity of the gas operative please contact Gas Safe Register on Tel: 0800 408 5500

Gas supply

- connect all appliances covered in this Part to the gas supply by a permanently fixed rigid pipe
- when you plan installing the gas supply for appliances in this Part, install the installation pipe according to manufacturer's instructions (see also the current Essential Gas Safety – Domestic – Part 5 for further guidance)
- the final connection to the appliance must incorporate an isolating tap and means of disconnection, to facilitate removal for servicing/maintenance etc.
- where the gas supply is to be taken through a wall, you must sleeve it
- where the gas supply is to be buried in the structure or run within the brick/masonry chimney recess, suitably protect the pipework from corrosion
  (for example, coat or wrap it with PVC tape)

System requirements

Open vented system

- fit a feed and expansion cistern to a boiler connected to an open system
- support the cistern on a flat, level, rigid platform capable of withstanding the weight of the cistern when filled with water
- the platform needs to support the whole base of the cistern and extend outward to a minimum of 150mm on all sides
- locate the cistern at least 1m above the highest point of the circulating system, or at such lesser height as specified in the boiler and circulating pump manufacturer's installation instructions. The cistern should have a capacity of at least 18 litres
- provide an open vent pipe from the circulating system to discharge over the feed and expansion cistern above the level of the overflow connection. It must not be less than 22mm diameter, should rise continuously and you must connect it in such a position as to prevent discharge of water or ingress of air in all normal conditions of service
- there should not normally be valves or components other than full bore pipe fittings between the boiler and the discharge point of the open vent
  (see Note)

Note: Unless a boiler manufacturer's installation instructions specifically state otherwise, the open vent and cold feed may be combined. The system doesn't need to have a separate open vent, only a cold feed and expansion cistern to accommodate hot, expanded water.

When you use this method, the boiler manufacturer's instructions often stipulate that the combined cold feed and open vent must not be less than 22mm in diameter.

When to use a close-coupled open vent and cold feed arrangement

Appliance manufacturers often specify this arrangement where:

- it is not possible to take the cold feed and open vent pipes back to the boiler; or
- where you are installing a 'low head' installation

Generally, both the open vent and cold feed connections into the system are made into the flow pipe from the boiler. In this arrangement, the circulating pump is also fitted in the flow pipe, but after the open vent and cold feed.

The general layout for this arrangement is as follows:

- position the open vent first in line at a maximum distance away from the boiler, which is defined in the manufacturer's instructions
- site the cold feed connection downstream of the open vent connection, but within 150mm of the open vent
- then site the circulating pump a short distance downstream of the cold feed connection. With this arrangement, the short distance between the cold feed connection and the inlet of the circulating pump is at a negative pressure, whereas the rest of the system is under a positive pressure

A distance in excess of 150mm is likely to result in a differential in system pressure between the open vent and cold feed connection. This differential may result in water being discharged from the open vent into the feed and expansion cistern during initial start-up of the circulating pump. Once this occurs water may continue to flow from the open vent due to the syphon effect.

A system affected this way is likely to corrode severely - as system water impregnated with dissolved oxygen is continually entering it.

Note: Ensure the system control arrangement doesn't create a closed off route back to the appliance if the boiler thermostat malfunctions – do this by including a system bypass arrangement (see 'System bypass' in this Part).

In some cases (particularly with low head installations where the ceiling height restricts the height of the open vent pipe) some manufacturers specify a 'surge arrester' in the open vent pipe.

- if so, you must make and install a larger diameter section of pipe (greater than 42mm in diameter) – to absorb the movement in the water levels, preventing pumping over (see Figure 8.13 and **System requirements** in this Part for further guidance)
- generally you do not need to fit a safety valve to an open system - unless the boiler manufacturer specifically requires this

Air separators – why they are used

One method of air removal you may find, is the fitting of an 'air separator' in the flow pipe from the boiler. This device provides a connection onto the open vent. It may also include a connection for the cold feed.

It is particularly useful during commissioning (or any subsequent re-commissioning) to quickly remove air from the system water.

The principle is that any air bubbles held within the heating system water are separated when reaching this device and allowed to vent off through the open vent pipe. As the volume of the air separator itself is greater than an equal length of flow pipework, the turbulence created causes the water to momentarily slow down - and allow any air in the water to separate and vent off via the open vent (see Figure 8.14).

Figure 8.13 A typical low head installation

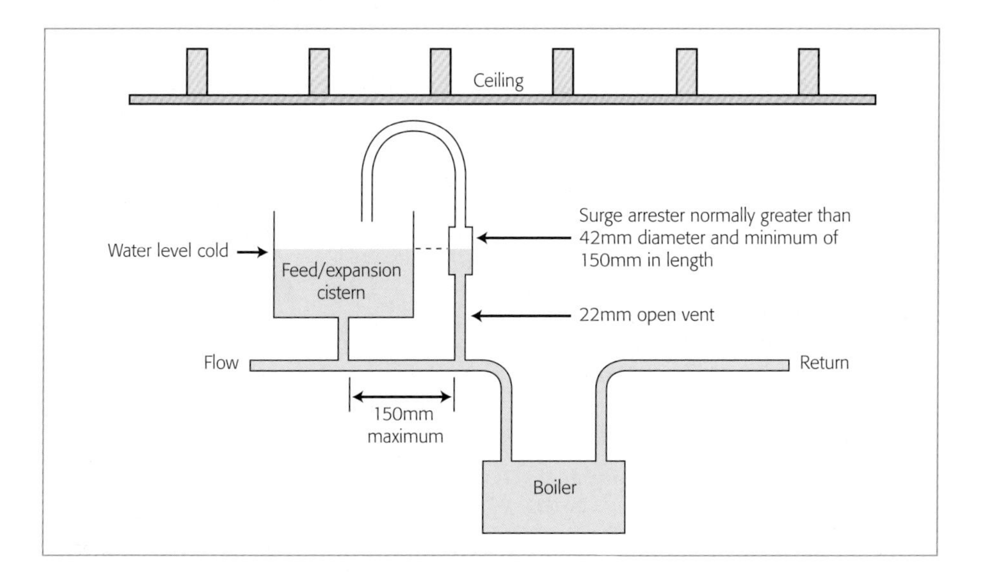

Figure 8.14 A typical water air separator

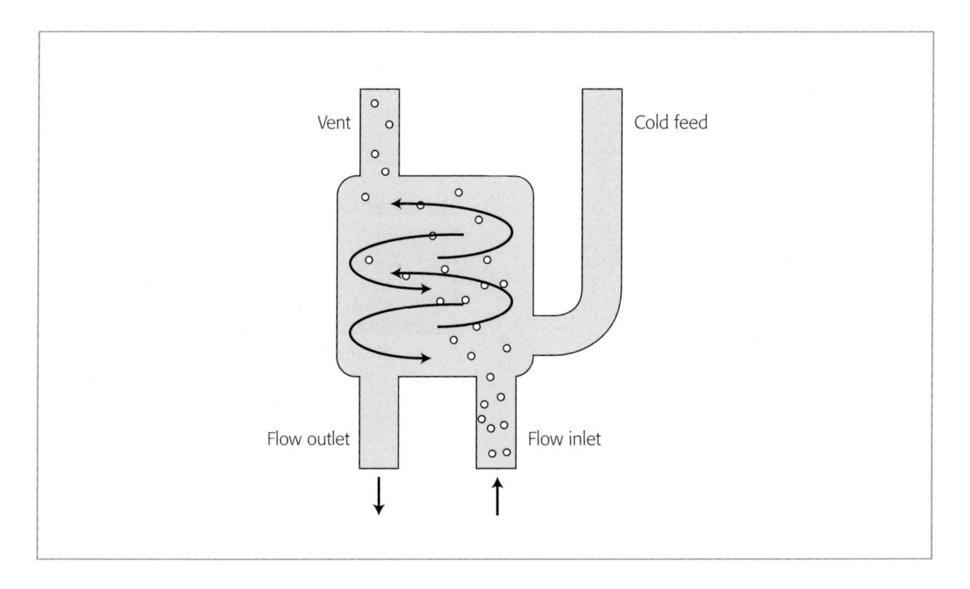

Figure 8.15 Typical sealed system components

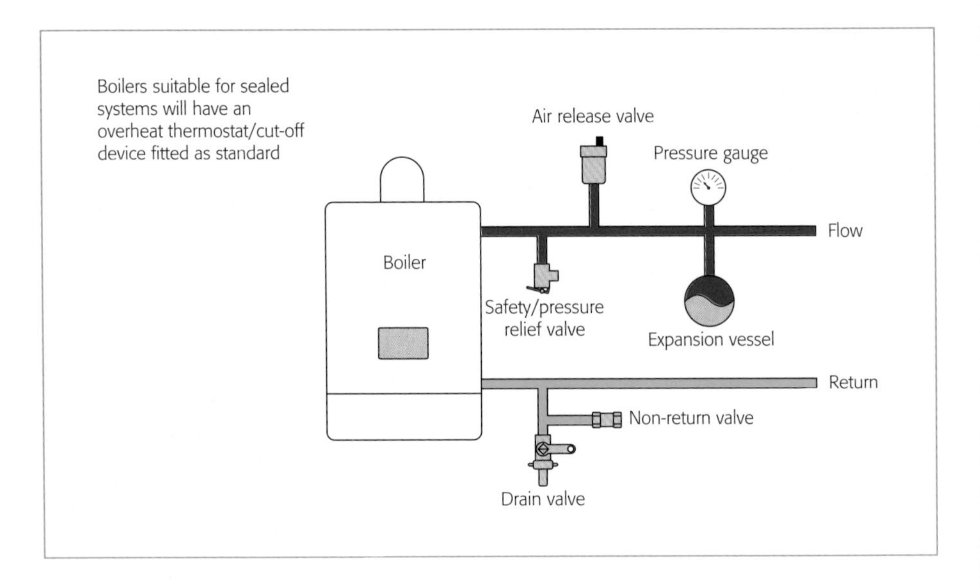

Sealed system

You are often required to convert an open vented system into a sealed system.

- if so, check the condition of the existing central heating system, (e.g. radiators, valves, etc). As these components may be very old and worn, they may be prone to leakage and damage when subjected to the higher pressure associated with a sealed system.

For example, sealed systems generally operate between 1 and 2 bar as opposed to less than 1 bar pressure in the average two-storey dwelling. Loss of pressure often results in boiler shutdown and can be due to leaking radiator valves.

- consider replacing these valves during the conversion process

Gas boilers can be purchased as a package suitable for installation to a sealed system - with the appropriate safety devices, components and controls fitted.

- if so, follow the manufacturer's installation instructions and connect the appropriate pipework to the boiler

However some manufacturers of 'standard' boilers claim that their boilers are suitable for sealed systems and provide the appropriate overheat device as an optional extra. Figure 8.15 identifies the typical sealed system components.

You will need to purchase the other components, controls and devices separately and fit them to the system.

In all cases in which you fit a boiler to a sealed system, it must have these essential components:

1. An expansion vessel to accept the expanded (heated) water (see **System expansion vessel – points to check** in this Part).
2. A non-adjustable safety valve (generally pre-set to release at 3 bar pressure).

Note: The safety valve discussed in this manual is in relation to the heating circuit and not one concerned with hot water storage vessels. For further information relating to hot water storage vessels refer to the current Gas Installer Manual Series – Domestic – 'Water Heaters'

3. An overheat thermostat acting directly on the gas valve.
4. A pressure gauge with a fill pressure indicator.
5. A filling point with a non-return valve facility.
6. An air release valve.
7. A drain valve.

Where you fit components 1 and 2 separately, it is important there are no shut-off valves or restrictions between the expansion vessel and boiler and the safety valve and boiler.

System expansion vessel – points to check

The majority of combination boilers installed are fitted to sealed systems. So there is no expansion cistern to accept the expanded system water once it is heated. With a sealed system, the expansion of water is accommodated by providing an expansion vessel. This is generally a component part of the boiler.

- if the expansion vessel is not a component part of the boiler and is located in a position remote from the heating circuit, the connecting pipe between the expansion vessel and the system pipework must have an internal diameter of not less than 13mm
- make sure its connection with the circuit is at a point close to the circulating pump inlet - in order to maintain positive pressures throughout the system
- the expansion vessel must have a capacity to accept the expansion of the system water when heated from 10°C to 110°C, without raising the pressure in the system to within 0.35bar below the lift pressure of the safety valve
- when you design a sealed heating system or install a replacement boiler, you must check that the capacity of the expansion vessel will accept expansion of the system water when heated to 110°C

In the majority of domestic systems the vessel provided by the boiler manufacturer should be adequate. If it is too small, the extra expanding water will be forced out of the safety valve or 'pressure relief valve'. This water will be lost and, unless an automatic means is provided for replenishing this water, the cycle will continue - until there is insufficient pressure remaining within the system for the boiler to remain operational.

- if this is the case, add an additional expansion vessel to the system

Information on how to calculate expansion vessel sizes is given in BS EN 12828: 2003 'Heating systems in buildings. Design for water-based heating systems'.

Note: BS EN 12828 replaces the previous standards – BS 5449: 1990.

Figure 8.16 Method of filling a sealed system

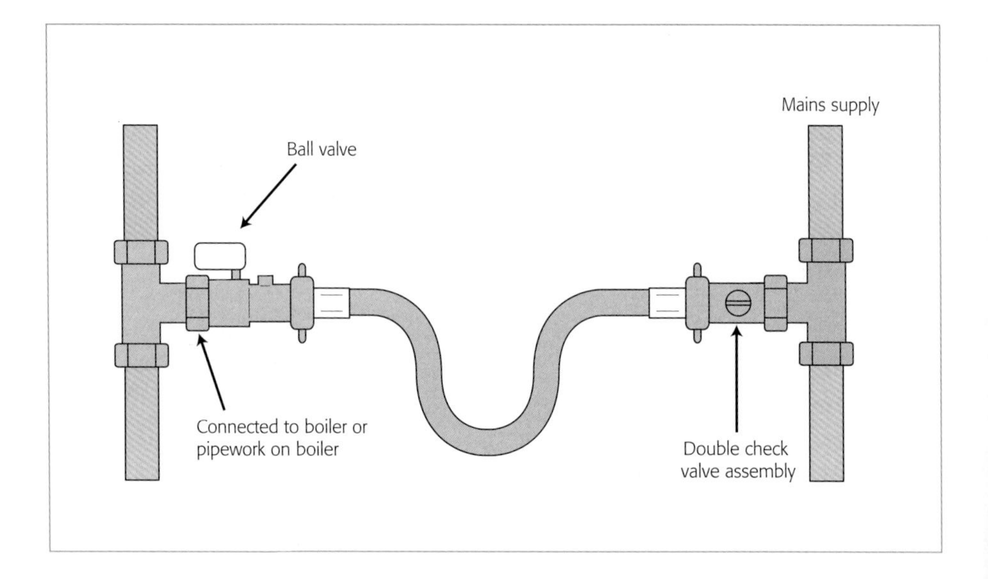

Filling point

When you fill or 'pre-pressurise' a sealed system, you may use the mains water supply - providing the connection you make complies with the Water Supply (Water Fittings) Regulations. These require that no supply pipe or secondary circuit should be permanently connected to a closed circuit for filling a heating system, unless it incorporates a back-flow prevention device in accordance with an approved specification.

The regulation is normally satisfied by using a temporary connection, provided that:

1. The connection is made through a double check valve assembly or some other no less effective device permanently connected to that circuit; and
2. You remove the temporary connection after use. Figure 8.16 illustrates a typical method of filling a sealed system.

Overheat thermostat

The overheat device will be a component part of the boiler, or it may be a manufacturer's accessory.

It generally takes the form of a thermocouple interrupter e.g. a heat sensitive device, fitted to the boiler flow pipe and connected electronically to the thermocouple. When activated, the device breaks the electrical current from the thermocouple to the multifunctional control valve, preventing gas passing to the burner.

Safety valve

- the safety valve is provided as a last line of defence - to release excess pressure from the system if all other controls fail

Although generally set to release at 3 bar, under fault conditions, any release of 'overheated' water from the valve will instantly turn to steam, increasing in volume approximately 1600 times.

Figure 8.17 Discharge point from a sealed system

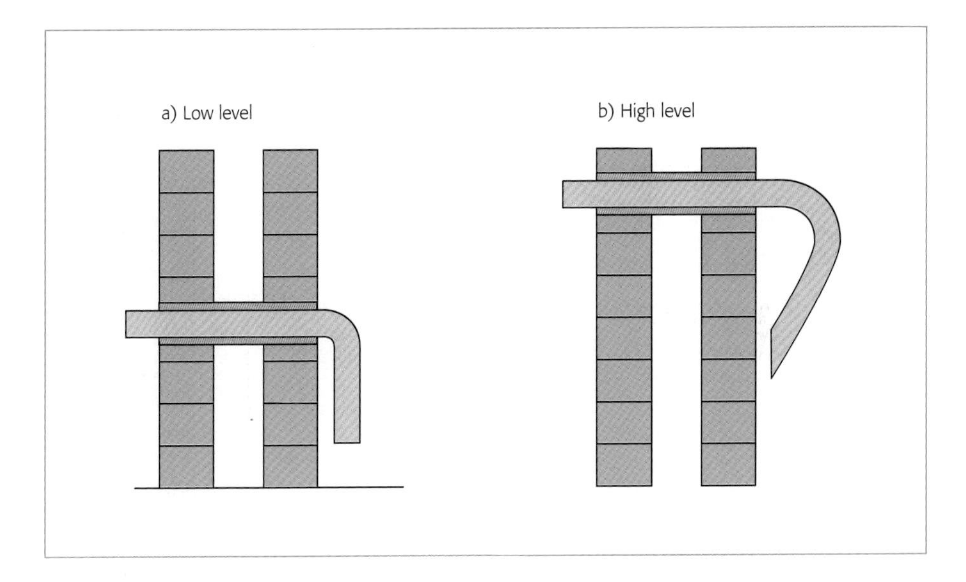

- so you must connect a discharge pipe to the valve, terminating in a position where it is clearly visible - and where any discharge cannot harm persons or property (see Figure 8.17)

Note: An additional figure, Figure 8.18 has also been included in this manual as additional good practice for discharge pipe termination, although its primary focus is the safe discharge of hot water from thermal stores covered in the current Gas Installer Manual Series – Domestic – 'Water heaters'.

- if the discharge pipe exits horizontally from the building it is recommended you face the discharge point downwards
- where the discharge is at low level, face the termination point downwards
- where the discharge is at high level, if you cannot terminate at low level, face the termination point downwards and back to the wall - to prevent scalding water coming out which could harm people or property. Alternatively, pipe cowl fittings and other proprietary devices are available that once installed will minimise the risk to persons below the point of discharge.
- the discharge pipe must be manufactured of a material that won't fail when subjected to high temperatures (e.g. copper pipe)
- install the discharge pipe so there is a continuous fall from the safety valve to the point of termination
- if you install the safety valve as a separate component, fit it in the flow pipe with no intervening valve or restriction, as close as possible to the top of the boiler

Attention: Any rise in the discharge pipe before termination could trap water. In wintry conditions this could freeze, forming an ice plug, rendering the safety device inoperative.

Figure 8.18 Discharge point from a sealed system

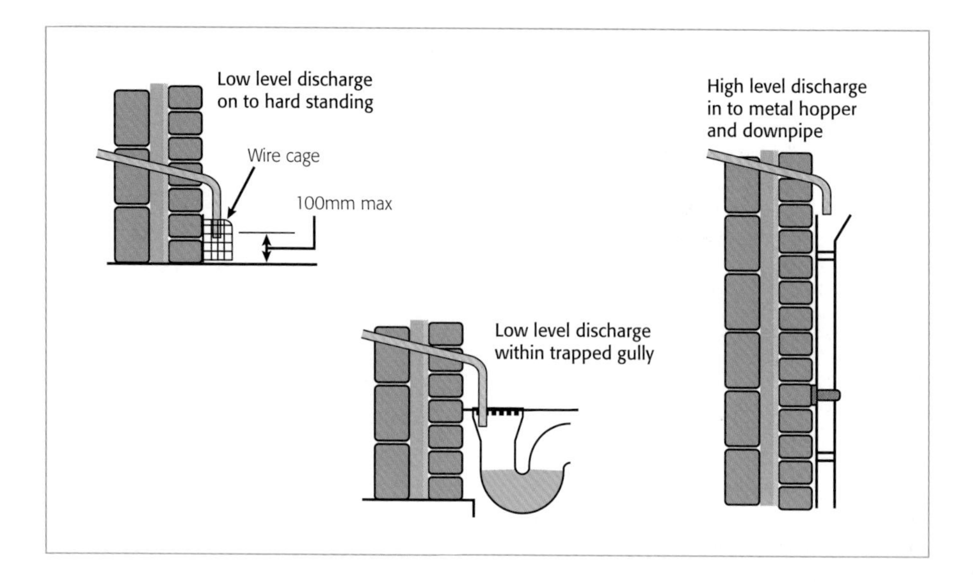

Pressure gauge with a fill pressure indicator

- permanently fix a pressure gauge with a fill pressure indicator to a sealed system
- ensure the gauge is easily seen from the filling point - and preferably connect it at the same point as the expansion vessel

Ventilation – essential for safe operating

All open-flued gas appliances need air for combustion and to assist the safe operation of the chimney system:

- gas appliances up to 7kW heat input (gross or net) generally do not require additional ventilation to be provided, as they rely on adventitious ventilation to the room for their correct operation - and that of the chimney (see Note)
- all ventilation requirements for gas appliances must follow the manufacturer's instructions. See also the current British Standard for ventilation requirements (BS 5440-2) for further guidance

Note: Modern construction techniques combined with better energy efficiency materials may mean that a property is very 'air tight' – air permeability less than or equal to $5m^3/h/m^2$ – negating the effect of adventitious ventilation.

In such instances (discussion with the building designer/developer will be required to ascertain if this is indeed the case), additional ventilation will need to be provided.

System water

Clean any existing system

Most central heating boilers are suitable for connection to an existing system. But to ensure that the new boiler has the best possible start and to reduce potential noise, clean the system with a proprietary central heating system cleanser (untreated systems can lead to accelerated wear, potentially reducing the operational lifespan of the appliance and its controls).

The cleaning process should be in accordance with the manufacturer's specifications and BS 7593: 2006 'Code of practice for treatment of water in domestic hot water central heating'.

Flush a new system

To ensure a central heating system or back circulator installation has the best possible start and to prolong its operational life:

- thoroughly flush the system out at least twice with clean water - ensuring that all the valves are open – once with cold water and once with hot water
- preferably disconnect the circulating pump and remove it from the system for the first flush out

The flushing process should remove all debris e.g. steel wool, slivers of copper, flux etc. from within the system.

Use proprietary cleaning agents to remove any remaining residues, such as flux and protect the system upon completion with an appropriate inhibitor complying with the manufacturer's specifications.

For further detailed guidance on cleansing central heating systems, refer to CORGI*directs* 'Cleansing domestic heating systems' pocket guide.

Warm air heating – 9

9 – Warm air heating

Figures

Tables

Introduction

This Part provides guidance for the installation, service, maintenance and replacement of warm air heating appliances that are 'CE' or British Standard (BS) kite-marked.

You may also refer to the information for used or second-hand appliances that do not carry the CE or BS mark – but have a data plate stating that the appliance is suitable for the gas type and pressure i.e. Natural gas at an appliance inlet pressure of 20mbar and Liquefied Petroleum Gas (LPG) Propane and Butane where the supply regulator has been set to provide an operating pressure of 37mbar and 28mbar respectively.

Note: Read this Part in conjunction with the relevant parts of the current Gas Installer Manual Series – Essential Gas Safety – Domestic – Parts 1-17.

How the systems have evolved

Like wet central heating, warm air heating systems supply heat from a central unit to a number of rooms or spaces – the main difference is that warm air systems use ducts and diffusers and/or registers to distribute the heat rather than pipes and radiators

Domestic warm air heating systems have been in use since the mid 1950s. Early heaters were inefficient as:

- they had small heat outputs and low fan power, combined with the fact that large outlets (registers) had to deliver heat into properties that generally had poor insulation with high heat losses
- they did not provide the comfort levels considered acceptable today. The belt-driven fan they used circulated warm air through a prefabricated duct system – and this usually only heated the living room, hall and kitchen areas

As the warm air heater market expanded, manufacturers improved comfort levels and efficiency with higher output heaters used with short, tailor-made ducts and small side wall registers.

Then in 1961 a Government report on minimum standards for housing included heating standards. It called for temperatures of 18°C in living rooms and 14°C in kitchens and other areas e.g. halls and landings – but did not consider that it was necessary to heat bedrooms and bathrooms.

Systems with low side wall registers could be installed cheaply – and local authorities and private builders fitted over a million in the 1960s and 1970s.

Other types of warm air heating systems also came in, such as the 'brick central' or natural convection system (see **Types and application of heaters – choose to suit dwelling** in this Part).

Some warm air heaters had water heaters (circulators) integrated into the design to provide domestic hot water. In many other dwellings a separate multi-point water heater was added.

Initially, all warm air heaters were of the open-flued type – but high-rise dwellings were followed by a design of room-sealed heaters that operated on 'Se-duct' and 'U-duct' chimney systems. Room-sealed free-standing models were also available
(see the current Essential Gas Safety – Domestic – Part 13 for further guidance).

The benefits of today's gas-fired warm air heater

It is more compact and fuel efficient than the earlier ones – and provides significant advantages for customers over wet systems, especially in modern highly insulated homes.

1. The freedom to site furniture almost anywhere in a room is very attractive - especially in smaller homes. Warm air diffusers and registers are much smaller than radiators and can provide an equivalent heat output. Modern fans, which are now directly driven (i.e. they have no fan belt) and sophisticated electronic heating controls deliver the air imperceptibly, to maintain a controlled temperature.
2. Fast response to controls, enables rooms to warm up quickly with the heater reaching operating temperature within minutes of being switched on and circulating the warmed air throughout the dwelling (a dwelling that heats up quickly ensures economy and comfort).
3. Warm air systems have the added advantage of improving the quality of air distributed around the home. Electrostatic air filters can be fitted to the heater to remove pollen and other irritants.
4. In a modern constructed home where draughts are greatly reduced, warm air provides a clean healthy environment and can introduce fresh air to reduce the possibility of condensation.
5. Some warm air heating systems have a facility for use in the summer months to provide fresh, filtered air throughout the dwelling. Some models have automatic humidity control.

Correct design and installation of warm air systems is essential

An efficient, effective heating system which meets customers' needs has to be:

- correctly designed; and
- correctly installed

General guidance on the design of warm air heating systems

- the information required must include detailed drawings of the building - showing the structure, building materials used and their U-values (to calculate heat losses in rooms/spaces)
- the customer's specification of internal design and desired temperatures is also useful
- this information will determine:
  - the amount of heat required for each room/space
  - the duct sizes for the air distribution and return air system; and
  - the correct size of the warm air heating appliance
- it helps to use a worksheet to assist you calculate logically

Note: Add 10% to the heat loss calculations for each room/space. This margin gives a reserve that allows for different wind conditions and ensures quick response to prolonged periods of cold weather.

Detail positions of the following:

- diffusers/registers
- air distribution/return air duct runs
- return air grilles, return air relief openings
- appliance position
- type and/or route of chimney system
- ventilation grilles

The design air temperature of a dwelling, in order to maintain comfort conditions for the occupants is the same for warm air systems as it is for wet central heating systems discussed in earlier Parts of this manual.

Refer to **Part 1 – Central heating wet – Comfort conditions – the basic requirements** for further guidance.

A correctly designed warm air heating system depends on two major factors:

1. Air temperature; and
2. Air movement.

- accurately sized ducts and warm air outlet registers/diffusers should achieve the correct air temperature
- correct air movement is achieved by air velocity through the outlets and the return air ducts/grilles being sized and adjusted to give the required air change in the rooms/spaces

Types and application of heaters – choose to suit dwelling

For effective operation, warm air heating systems depend on the continuous circulation of air, which is heated and distributed to one or more rooms, or internal spaces, simultaneously. Warm air heating appliances fall into two categories:

1. Fan assisted.
2. Natural convection.

Either of these may be combined with a water heating appliance (usually a circulator), to supply domestic hot water demand.

So when you consider an installation, you must select the type of appliance that suits the type and size of dwelling.

You will find further guidance on water heaters in the current Gas Installer Manual Series – Domestic – Water Heaters.

Fan assisted model

Warm air is fan assisted through a network of ducts. Displaced air is returned to the heater by means of a series of return air grilles/air relief openings connected by a return air duct to the warm air heater.

Preferably site the warm air heater centrally in the dwelling to keep duct lengths, including the return air duct, to a minimum.

Some models are room-sealed, but the majority of installations will be open-flued (due to the fact that they need to be centrally located).

Heaters are designed with open-flued connections that are located at the back, top rear or top front of the appliance (see Figure 9.1). They are most frequently fitted in cupboards or compartments. Some models utilise any convenient narrow space (slot-fix) or may be free-standing.

Fan assisted heaters are categorised with reference to the air flow movement through them.

They are generally divided into three main types (see Figure 9.2).

1. Up flow – generally free-standing, taking air from low level through a heat exchanger to discharge from high level ducting.
2. Down flow – generally free-standing, taking air from high level through a heat exchanger to discharge from low level ducting. This type is the most commonly used.
3. Horizontal (cross) flow – generally wall-mounted, taking air through a heat exchanger to discharge from side ducting.

Figure 9.1 Open-flued and room-sealed chimney connections

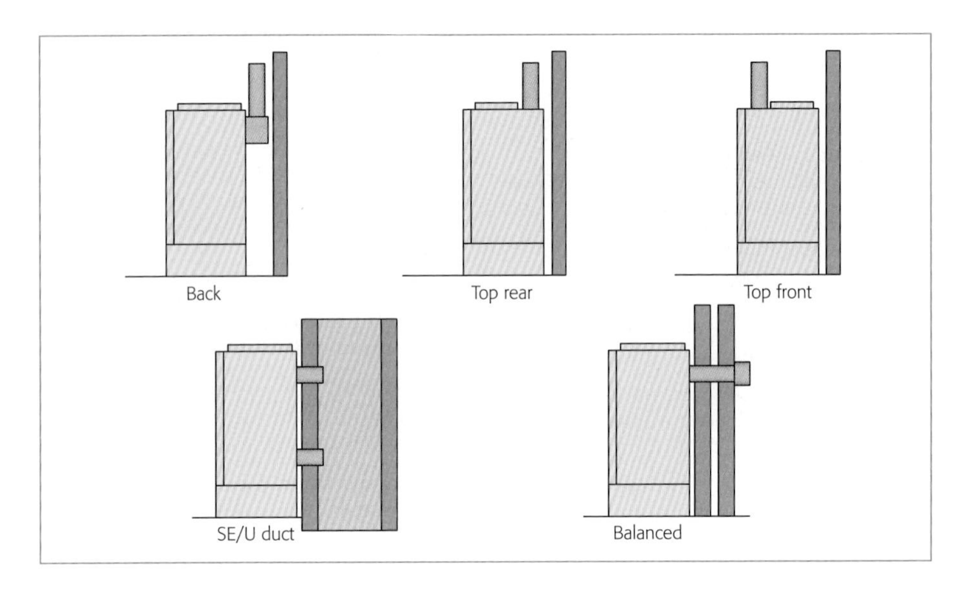

Figure 9.2 Basic types of fan assisted warm air heaters

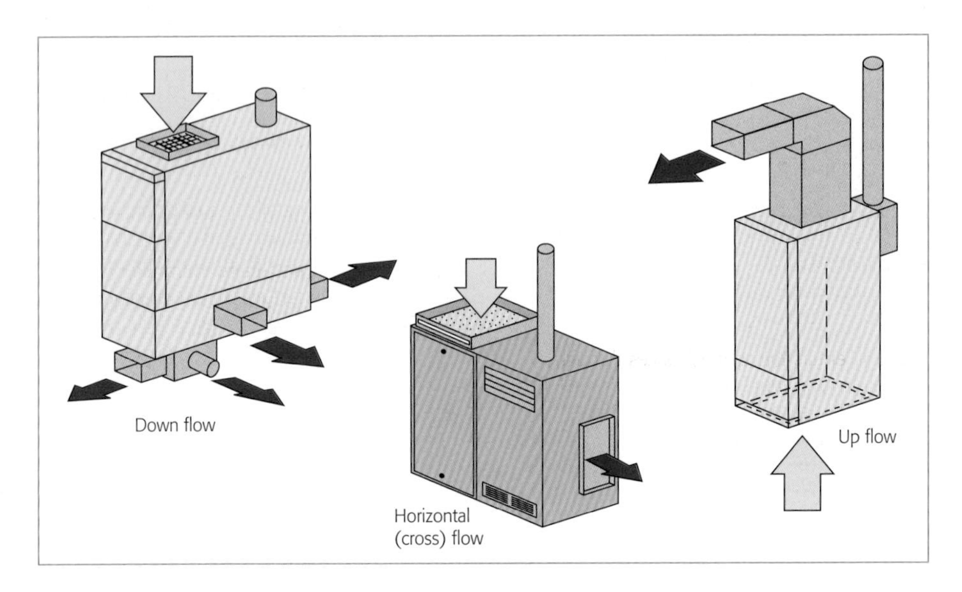

Figure 9.3 Typical downflow heater

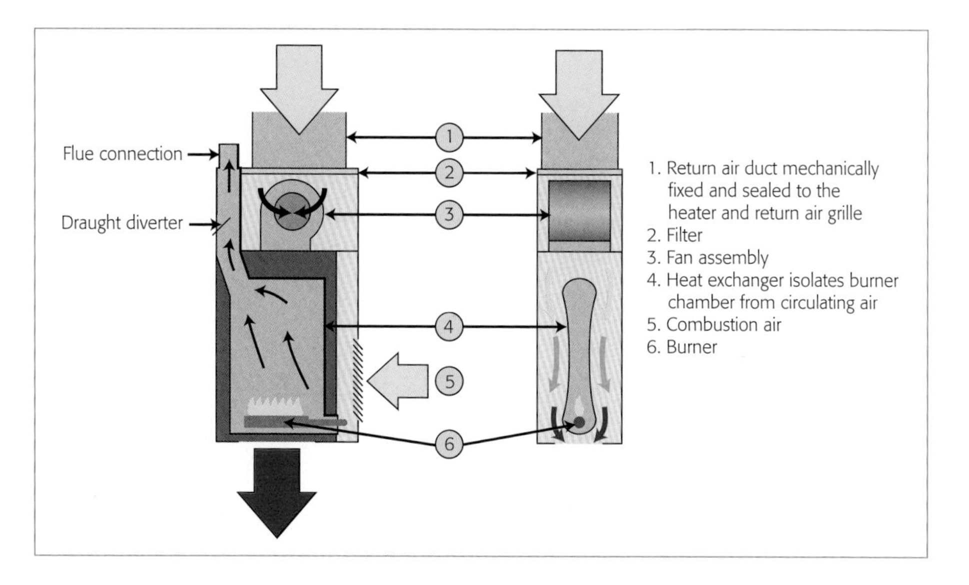

How it operates

The air to be heated is drawn into the heater via the return air grille - and through a filter or electronic cleaner (see **Return air (re-circulation) – Air filters for warm air heaters** in this Part).

The air is fan-assisted to pass over a heat exchanger where its temperature is raised by 50°C ±5°C. The warm air enters a plenum (see **Warm air plenum – needs to be sealed and bear weight** in this Part), which equalises the pressure and temperature before it is distributed into the dwelling via a network of ducts.

Warm air enters the rooms/spaces through diffusers or registers fitted on the side wall, floor or ceiling. Displaced cooler air is then returned to the heater (see **Return air (re-circulation)** in this Part) for re-heating.

A fan incorporated in the warm air heater provides the power for this circulation. Figure 9.3 shows the principle of operation of a typical down flow heater.

Note: Since 1969 open-flued domestic warm air heaters are required to have a connection made from the return air spigot on the heater to the main return air grille(s).

This positive return air connection prevents the circulation fan from interfering with the burner and an open-flued chimney system of the heater (see 'Return air (re-circulation)' in this Part).

Figure 9.4 Typical natural convection or brick central installation

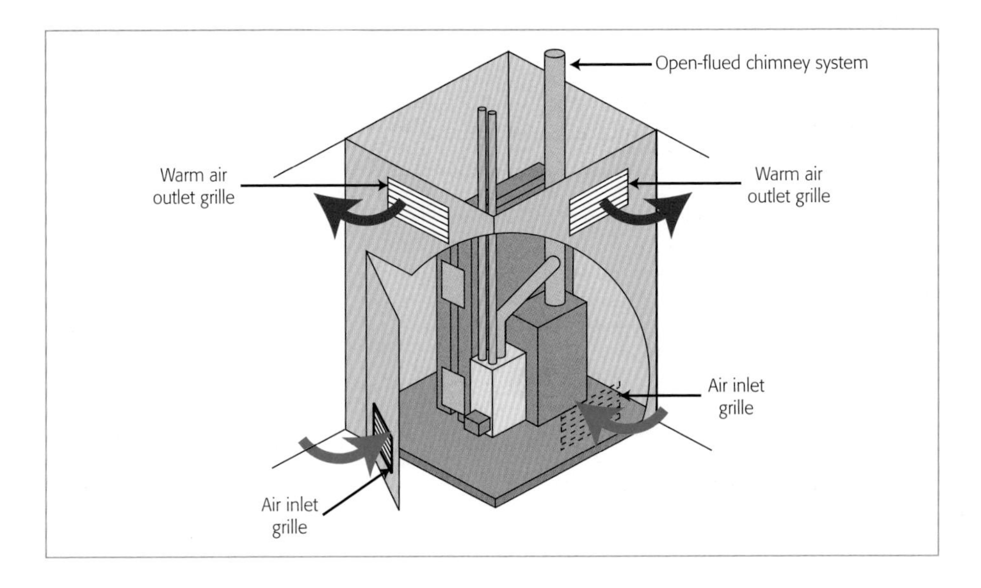

Natural convection model

- sometimes known as a 'Brick Central' model, air movement relies on natural convection to heat the dwelling. This type is not suitable for fan assisted ducted warm air
- the warm air heater is designed for you to install in a convection chamber, which is located in the centre of the dwelling. This is a purpose-made compartment, constructed of brick, block or prefabricated panels
- if a natural convection warm air heater is specifically designed for siting in a convection chamber, make sure the chamber complies with **Compartment installations – what you must do** in this Part
- also make sure that no air vent or duct communicates directly between the chamber and any floor(s) sited above the chamber

How it operates

Low level air inlet grilles allow cold air to enter the chamber, where it is warmed by the appliance heat exchanger. The warmed air is then discharged, by natural convection, through grilles or registers fitted at the top of the chamber (see Figure 9.4).

Due to the cooler air being heated by a heat exchanger, the appliance is generally not fitted with an outer casing.

Conversion to another gas

Carry out conversion to another gas, if necessary, strictly in accordance with the appliance manufacturer's instructions and using their supplied kit of parts.

Restricted locations – for safety reasons

Restricted locations for warm air heaters are the same for any open-flued or room-sealed gas appliance, and therefore, as previously discussed within this manual for boilers.

For open-flued, refer to **Part 2 – Open–flued boilers** and for room-sealed, **Part 3 – Room-sealed boilers; Restricted locations – for safety reasons**.

Installation – what you need to consider

To avoid repetition, general information will be found in **Part 11 – General installations details – Warm air systems**.

You can install warm air heaters in a number of ways. By far the most popular modern warm air heating appliance is the fan-assisted type – and this guidance applies to fan assisted heaters, but you must read it in conjunction with the manufacturer's installation instructions.

At your initial planning stage, consider the following:

1. Note the size, type and location of the warm air heater. Allow adequate space for the heater, its compartment or recess, plus the clearances and access space recommended by the manufacturer (by placing the heater centrally, you can keep duct runs short).
2. Keep associated ductwork (warm air and return air) system connections as short as possible. Short duct lengths increase efficiency and reduce costs. Where possible, take the main return air duct from the hall or landing.
3. Position warm air outlets (diffusers/registers) to give optimum comfort levels.
4. Note the number of storeys in the building - as this may determine additional requirements.
5. Note the chimney system type to use (room-sealed or open-flued).
6. Note the location and/or route of the chimney system. For open-flued heaters, a simple vertical path will generally be suitable, as most heaters will be positioned centrally in the dwelling.
7. Consider the ventilation requirements (see **Part 11 – General installation details – Warm air systems – Ventilation – when and why needed**).
8. Consider which gas and electric supplies are available/suitable.
9. Where applicable, note the position of the domestic hot water storage vessel.

Types of buildings

Buildings of one or two storeys

Install warm air heaters in buildings of one or two storeys following the manufacturer's installation instructions.

Buildings of more than two storeys

Building Regulations (ADB Fire Safety – Volume 1 Dwellinghouses – England and Wales) specify additional requirements you must comply with when you install a warm air heater in any flat or maisonette which has one of its floors, situated 4.5m or more above access level.

This is because in the event of a fire, a warm air heating system could allow smoke and fire from the dwelling to spread into a protected entrance hall or landing.

The following requirements also apply to any flat or maisonette in blocks of more than two storeys - with the exception of one and two storey dwellings, located within the blocks, which have their own access at ground level.

1. Do not fit air transfer grilles in any door, wall, floor or ceiling that communicates with a protected entrance hall or protected stairway.
2. Fit all ductwork passing through any wall, floor or ceiling bounding a protected entrance hall or protected stairway so that all joints between the ductwork and the boundary are fire stopped.
3. Where a duct conveys warm air into a protected entrance hall or protected stairway through any wall, floor or ceiling bounding the protected entrance hall or protected stairway, duct the return air from that area back to the heater.
4. Position warm air and return air grilles at a height not exceeding 450mm above ground level to restrict the spread of smoke.
5. Mount a room thermostat in the living room at a height between 1200mm and 1500mm above floor level.

Note: See also 'Restricted locations – for safety reasons – Protected shafts/stairway' in Parts 2 and 3, as appropriate.

Compartment installations – what you must do

You may install warm air heaters in purpose-made compartments (probably the most commonly used installation method) see Figure 9.5 – the compartment must comply with the following requirements:

1. It should be a fixed rigid construction.
2. If the appliance manufacturer's installation instructions have no specific advice, any internal surface of the compartment constructed of combustible material must be at least 75mm away from any part of the heater. Alternatively, line the surface with non-combustible material with a fire resistance of not less than 30 minutes. Materials complying with BS 476 generally meet these requirements.
3. Fit with air vents at high and low level for the provision of compartment cooling and if you install an open-flued appliance, also provide combustion air. Size the air vents following the manufacturer's instructions.

   See also the current British Standard for ventilation requirements: BS 5440-2 for further guidance.
4. Fit an access door to permit inspection, servicing and maintenance of the heater and any ancillary controls. The access door must be large enough to allow the heater to be removed.
5. To discourage its use as a storage cupboard, fix a notice in a prominent position, to warn the customer against such use (see also the current Essential Gas Safety – Domestic – Part 10 for further guidance).
6. Where the heater is open-flued:
   - connect the return air grille directly to the appliance by means of a continuous duct (see **Return air (re-circulation)** in this Part)
   - the compartment door or air vents must not communicate with a room containing a bath or shower
   - for installations where the airing cupboard door or air vents communicate with a bedroom/bedsitting room (see **Bedroom/bedsitting rooms** in this Part)
7. Air vents must not penetrate a protected shaft or stairway (see **Part 2 – Open-flued boilers** and **Part 3 – Room-sealed boilers – Restricted locations – for safety reasons**).

Figure 9.5 Typical compartment installation of a downflow fan-assisted ducted warm air heater

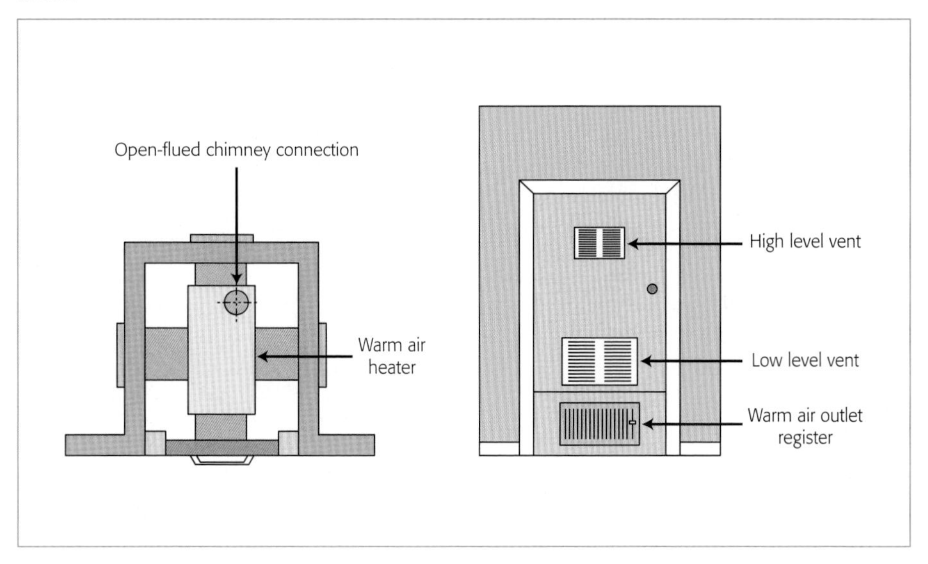

'Slot-fix' installations

These are suitable for open-flued, low and moderate output heaters. They use little floor space and you can site them on the ground or first floor. The 'slot' protects both sides of the heater.

The 'slot-fix' type warm air heater is for installation in a narrow recess with the front panel and controls exposed. Extra decorative panels fixed to the top of the heater fill in the opening to ceiling level, to prevent the draught diverter and/or combustion air vents and/or return air grilles from being obstructed by the occupants of the dwelling.

Combustion air enters the heater through the exposed panels or through special ventilation grilles incorporated into the design.

Note: You can only fit appliances specifically designed for 'slot-fix' application this way.

- take care to ensure adequate clearances from combustible materials
- all installations must comply with the manufacturer's installation instructions (also see image 'a' in Figure 9.6)

Free-standing/storey height installations

- in this situation, you install the warm air heater in a room or hallway with at least one side (preferably rear) against a wall (see image 'b' in Figure 9.6). A set of additional decorative panels extends the profile of the heater to ceiling level
- these models take minimum floor space
- generally the heater is on display - so its aesthetic appearance is important
- take care when you site the heater to avoid sound transmission from the heater fan
- also ensure adequate clearances from combustible materials

Figure 9.6 Slot fix and free-standing/storey height installations

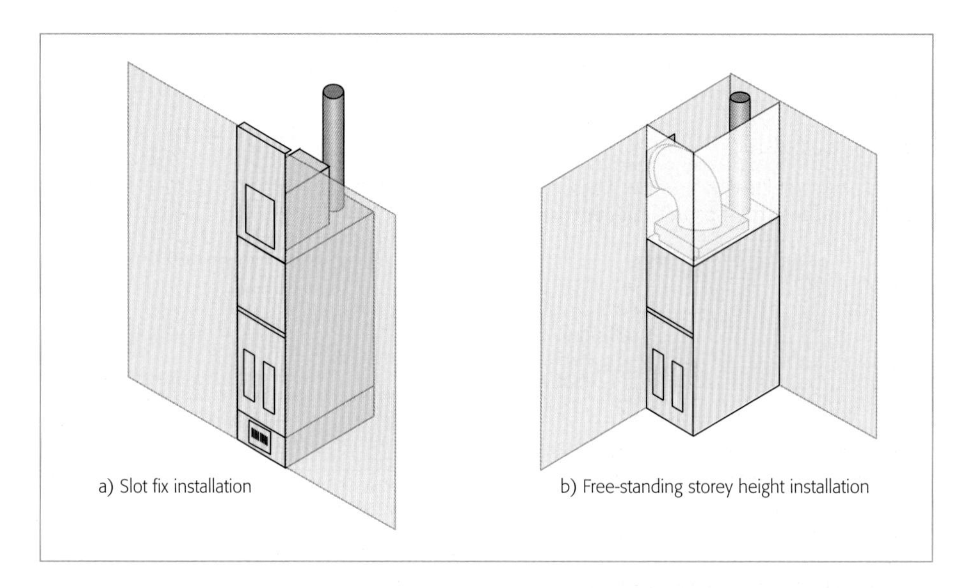

Airing cupboard installations – what to do

When you adapt an airing cupboard to house a warm air heater, follow the requirements in **Compartment installations – what you must do** in this Part.

Additionally, ensure that you:

- separate any other space(s) within the compartment from the heater by a non-combustible partition. If the partition is made of perforated material, the perforations must not exceed 13mm. This is to stop combustible material (e.g. clothes) from coming into contact with the heater or its chimney system
- where the heater is open-flued, locate the draught diverter and the air vents within the heater space
- separate the chimney system from the airing space with a non-combustible guard in order to prevent damage to the contents of the airing space

Note: For double-wall flue pipe conforming to BS EN 1856-1, the level of insulation provided by the air gap between the outer and inner pipe may be deemed sufficient to provide the necessary protection.

- if the flue pipe is single-wall, fit a non-combustible guard around the pipe, with a minimum of 25mm between the pipe and the guard

Clearances between the chimney guard and compartment partition must not exceed 13mm. Expanded metal or rigid wire mesh with apertures not exceeding 13mm are suitable materials for the partition and guard.

When to consider under stairs cupboard installations

If there's no practicable alternative location to site the heater, you can consider an under stairs installation. Whenever possible, the heater should be room-sealed.

A warm air heater fitted in an under stairs cupboard or space must comply with the following requirements:

1. If the premises in which the under stairs cupboard is located is no more than two storeys, then the cupboard itself must comply with the requirements in **Compartment installations – what you must do** in this Part.
2. If the premises in which the heater is located is more than two storeys, all internal surfaces of the cupboard, including the base, must be constructed of materials that are non-combustible, or lined with non-combustible material (see Note 1).
3. The air vents must communicate direct to outside (see Note 2). Size them following the manufacturer's instructions (see also the current British Standard for ventilation requirements: BS 5440-2 for further guidance).

Note 1: Non-combustible materials must have a fire resistance of not less than 30 minutes in accordance with the relevant part of BS 476.

Note 2: Under certain circumstances it may be possible to use intumescent air vents communicating internally – buildings greater than 3 storeys – but approval of the local BCB will be required before proceeding with the installation.

Roof space installations

Install warm air heaters in a roof space in accordance with the manufacturer's instructions, the general installation requirements are:

1. Provide a suitable flooring area capable of supporting the heater, duct work and any associated controls and equipment under and around the heater. If the flooring is of combustible material, provide a 12mm thick insulating base extending at least 25mm beyond the heater edges.
2. Make sure the heater is accessible for service and maintenance.
3. Provide a permanent means of access to the roof space (a fixed retractable roof ladder would be suitable).
4. Provide a safety barrier around the roof space opening.
5. Provide fixed lighting for the heater and access area.
6. Fit a safety barrier around the heater to prevent contact with any stored items in the roof space.
7. If the heater is open-flued, provide adequate ventilation in accordance with the manufacturer's instructions in the roof space.

Figure 9.7 Se-duct and U-duct installations

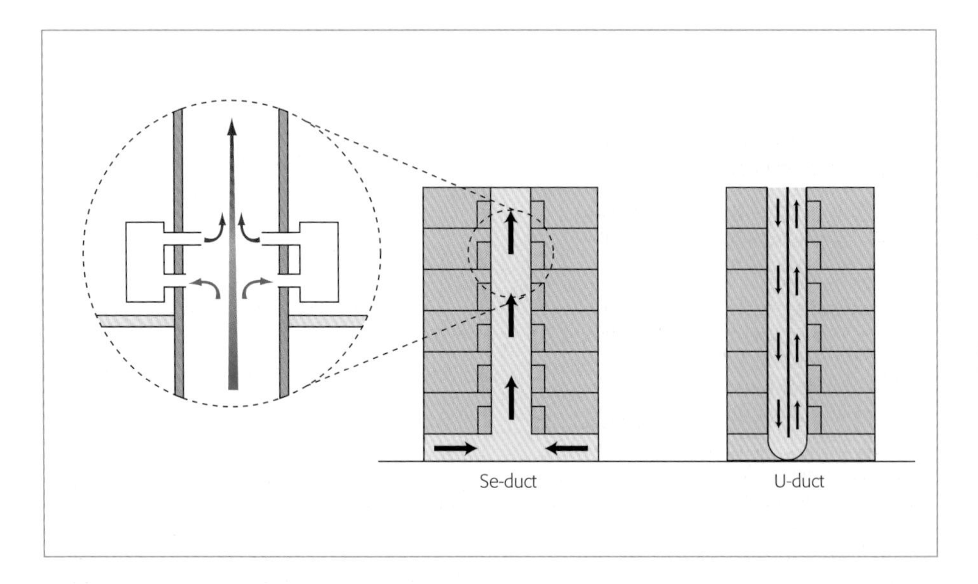

Se-duct and U-duct appliance installations

Special room-sealed appliances are designed to suit the Se-duct and U-duct systems. Ensure that the appliance is suitable for the purpose before you start installing.

Se-duct and U-ducts are shared chimney systems generally built into blocks of flats, usually high rise. Combustion air is taken from the shared chimney and POC are channelled back into the shared chimney system.

Se-ducts are designed so that the chimney system is open to atmosphere at both ends, allowing airflow through the duct. The air inlets (of which there is often more than one per duct) are positioned at low level and the termination or exhaust (of which there is only one per duct) at high level.

U-ducts are similar to Se-ducts in their function. The duct is arranged in a 'U' shape with both the air entry and termination (or exhaust) open to atmosphere on the top of the building. The inlet and outlet spigots of the appliances are connected to the up-flow leg of the system.

Special room-sealed appliances are installed against the ducts and connected to it by two open spigots in accordance with manufacturer's instructions. Each warm air heater takes in combustion air via its bottom spigot, which is normally flush with the Se/U-duct inner wall. The POC are discharged back into the same duct through the longer upper spigot (see Figure 9.7).

- you generally fit the warm air heater in a compartment
- construction and ventilation requirements are as discussed in **Compartment installations – what you must do** in this Part

Return air (re-circulation)

Since 1969, Building Regulations require you to fit domestic open-flued warm air heaters with a full and unobstructed return air arrangement. This prevents the circulation fan from interfering with the burner and chimney system of the heater.

- make a positive return air connection between the return air spigot on the heater and the main return air grille(s)

This requirement applies directly to open-flued appliances you install in a compartment – as well as 'slot-fix' and free standing storey height installations

In general terms, to ensure a safe installation for open-flued domestic installations in all situations/locations, it is recommended you make a positive return air connection between the return air spigot on the heater and the main return air grille(s)

- if you encounter an existing domestic open-flued warm air heater with fanned warm air circulation that does not have a positive return air arrangement – then (whenever possible or practicable) make or recommend a positive return air connection
- before you undertake any remedial work, consult the appliance manufacturers for the particular requirements for each appliance type

Attention: Where there is no positive return air connection in an existing installation, class the installation as 'At Risk' (AR) in accordance with the current Gas Industry Unsafe Situations Procedure (see also Essential Gas Safety – Domestic – Parts 8 and 10 for further guidance), but if there is no provision on the appliance to install a positive return air connection, you must seek advice from the appliance manufacturer, or other warm air specialist.

- provide a full and unobstructed return air path to the heater from all heated rooms and spaces – except kitchens, bathrooms and toilets. This allows air from heated rooms/spaces to be returned (transferred) to the heater for re-heating and re-circulation. The return air path itself is provided by return air grilles connected by ducts to the heater
- air relief openings allow air to move to the return air grilles from other rooms/spaces - do not fit these directly between bedrooms
- make sure the chimney system does not run through an area serving as a return air path

Return air grilles

Ideally, position return air grilles in hallways, landings or in another suitable 'collection area' within easy location of the warm air heater see Figure 9.8.

- connect the grille(s) to the heater by a duct. The collection area may have air relief openings from other heated rooms/spaces. Smaller dwellings require a single return air grille. In larger dwellings, several return air grilles may be required for each floor or area
- if you fit more than one return air grille, connect all grilles directly to the heater by using the same duct see Figure 9.8
- where a living room is used as the collection area, take care when you position the return air grille, to ensure that the cooler air moving from other rooms does not create draughts. Larger rooms may benefit from an additional return air grille serving only that room

Note: A collection area is where all heated rooms/spaces are connected to one area by a number of air relief openings.

Figure 9.8 Air relief openings

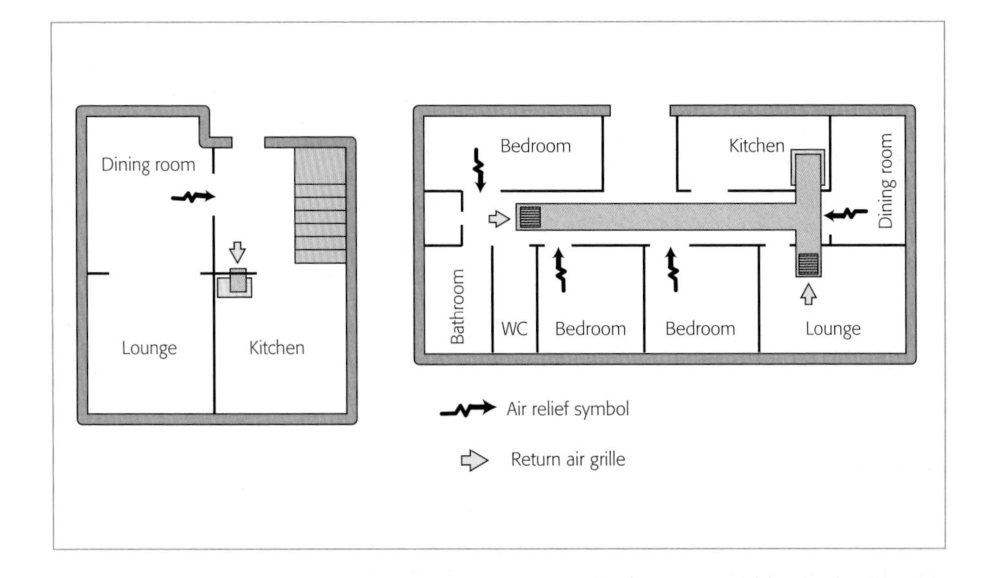

Air relief (or transfer) openings

These openings generally depend on the construction and layout of the dwelling (see Figure 9.8).

- you must fit them in all heated rooms (except kitchens, bathrooms and toilets) to provide a path to the collection area
- openings can be in doors or walls, but must not be subject to accidental blockage. Therefore, gaps under doors are not acceptable
- ensure that a grille covers the openings

How to position return air grilles/air relief openings

Return air grilles and air relief openings are designed to circulate the air within the room and return the cooler air back to the heater.

Consider these points in order to decide their positions:

1. If there's no alternative, you can use a living room as a 'collection area' - but take care to locate return air grilles and relief air openings - to ensure that cooler air returning from other parts of the building does not create draughts.
2. Preferably don't position return air grilles in a room containing an open-flued appliance (this includes solid fuel, oil and gas appliance types). If you can't avoid this, ensure the inlet velocity does not exceed 1.5m/s - and correctly size any air relief opening to the living room (see Attention).

3. Unless otherwise advised by the appliance manufacturer, you may site return air grilles at high or low level in walls, ceilings or floors (see also **Types of buildings** in this Part).
4. If the warm air distribution supply register/diffuser is at one end, position the return air grille or air relief opening at the opposite end of the room (particularly in long narrow rooms).
5. Position grilles at low level if warm air is supplied at high level and vice versa.
6. Don't site grilles adjacent to warm air outlets (they may cause the heater to cycle more frequently).
7. Avoid siting low level grilles in a likely sitting area (occupants may suffer from cold ankles).

Attention: If you don't make necessary allowances for return air grilles/air relief openings, a room in which a return air grille is sited and from which return air is taken, could be subjected to sub-atmospheric pressure (caused by the suction of the warm air circulation fan) – this could adversely affect the safe operation of the warm air heater itself - and/or any other open-flued appliance in the same or adjoining rooms.

Return air ducting

Open-flued appliances – what to do

- position the main return air grille(s) in a suitable 'collection area' e.g. a hall or landing. For ease of installation, generally fit them in a wall or a ceiling
- connect a duct from the return air grille(s) to the return air inlet on the appliance (appliances that have a choice of top and side return air connections make it easier to install this positive return air link)
- seal the duct so that the return air is separated from the combustion air within the heater compartment, or heater case. If you don't make this seal, it could adversely affect the open-flued chimney system operating safely
- the optimum method is to use a rigid metal duct (see Figure 9.9), or flexible return air kits which are simple to fit (see Figure 9.10)
- you must correctly fix the duct, secure it with screws or rivets to the wall/ceiling/heater - and seal it
- if you use a flexible duct kit, you may need to connect it to the square or rectangular return air connection on the heater. You can get an adapter for this purpose from the warm air heater manufacturer

Note: Make sure the flexible ducting doesn't come into contact with the hot surfaces of the chimney system or draught diverter of the warm air heater.

- design return air ducts to avoid transmission of noise from the warm air heater fan compartment. This is particularly important if you site return air grilles in living rooms, or where the heater has a relatively high fan speed
- to avoid this, don't use short return air ducts. Alternatively, incorporate at least one bend in the duct. If you can't do this, line short duct lengths with a sound absorbing material

How to size air relief openings/return air grilles/return air ducts

- size return air grilles and air relief openings to handle air equal in volume to the warm air supplied to the rooms/areas they serve. Size them similarly for open-flued or room-sealed appliances
- ensure the velocity of air through the grilles does not exceed 2m/s

Figure 9.9 Rigid metal return air duct

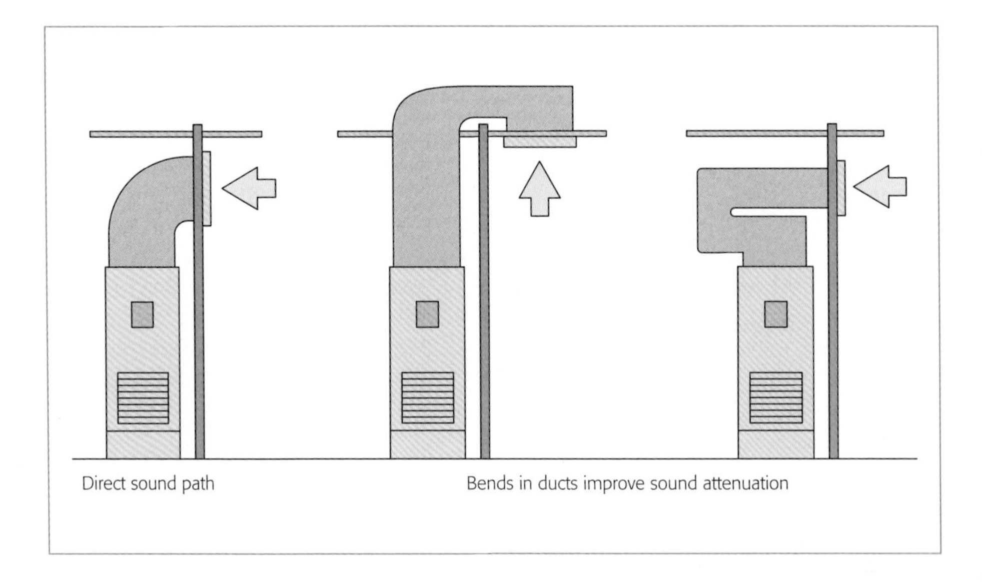

Figure 9.10 Flexible return air duct

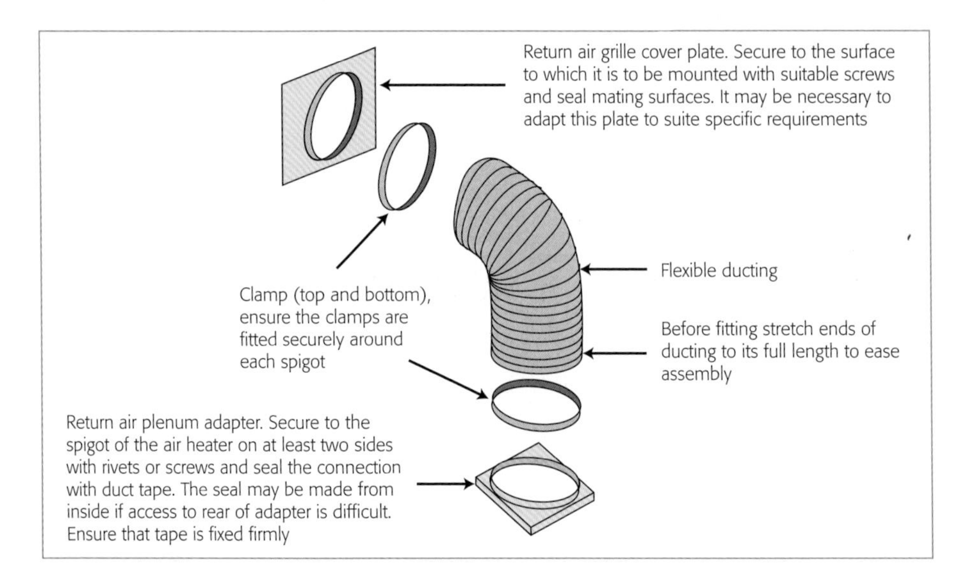

Table 9.1 Return air grille/duct sizing

Output up to kW	Return air grille size (mm)	Return air duct size rigid (mm)	Return air duct size flexible (mm)
7.5	350 x 250	200 x 200	250 diameter
8.8	350 x 300	250 x 200	250 diameter
10.3	350 x 350	250 x 250	300 diameter
11.7	400 x 350	250 x 250	300 diameter
13.2	400 x 400	300 x 250	300 diameter
14.7	450 x 400	300 x 250	300 diameter
17.6	450 x 450	300 x 300	350 diameter
19.1	500 x 450	350 x 300	400 diameter

- relief air openings should have a free area of 88cm^2/kW of heat input to a room/space

You will find a guide to sizing return air grilles and return air ducts in Table 9.1. It's based on typical grilles with a free area of 70% of the total grille size, which is a general manufacturing design. For grilles with different free areas, consult the manufacturer's data.

Attention: Don't underestimate the importance of providing correctly sized return air grilles/air relief (or transfer) openings.

If you fail to make necessary allowances for return air grilles/air relief openings, a room in which a return air grille is sited and from which return air is taken, could be subjected to sub-atmospheric pressure (caused by the suction of the warm air circulation fan).

This could adversely affect the safe operation of the warm air heater itself - and/or any other open-flued appliance in the same or adjoining rooms.

Whenever you install an open-flued heater, you must carry out a spillage test on the heater chimney system - to ensure no spillage of POC is caused when the warm air circulation fan is operating

- in cases where you detect spillage, check and re-check your calculations to ensure there's an adequately sized return air path back to the heater
- in the case of an open-flued heater installed in a location other than a compartment e.g. a room, you may need to install a positive return air connection between the return air spigot on the heater and the main return air grille(s) into the room

Room-sealed appliances

- you may install room-sealed appliances without a duct connecting the return air grille(s) to the return air inlet on the appliance - providing there's a return air path to the heater
- it is essential you ensure that the cooler air from other heated rooms returning to the collection area has an unobstructed return air path to the heater

A range of air filters for warm air heaters

Air filters come in a range of designs and materials. They are located in the appliance to filter the incoming air returning to the appliance and are generally:

1. A glass fibre blanket material, which is placed over an open cage in the form of a 'hammock' to present a large area to the incoming return air (used in both up-flow and down-flow heaters).
2. A fine plastic mesh which is held in place within a ridged frame, allowing for easy removal and cleaning (often used in slot-fix appliances).
3. Electrostatic/electronic air filtration – a method where even minute dust particles, tobacco smoke, pollen and other atmospheric pollutants are removed from the air by polarising the filter medium with an electrostatic charge (an ideal method of filtration for those persons who suffer from air borne allergies).

Generally, air filters should be removed and cleaned every two weeks during the heating season.

Duct systems – points to consider

The following guidance is based on the use of galvanised steel ductwork, which is the most commonly used. You need to consider any other material used to construct the ductwork individually, as to its suitability in relation to fire resistance and airflow.

Duct layouts and sizes are determined by their length and the volume of air (heat), which they carry. The most important factor to consider is the resistance to airflow.

Follow these guidelines:

1. Use the least number of duct fittings i.e. bends and tees.
2. Select components of low resistance (avoid high resistance components e.g. square bends without internal air guides, or sudden changes in the duct cross-sectional area).
3. Always use take-off fittings (branch ducts) that are angled in the direction of the airflow.
4. Position take-off fittings or branches on straight sections of the main duct. Don't connect them to elbows, bends or transition fittings.

Warm air plenum – needs to be sealed and bear weight

A warm air plenum is a 'box' designed to equalise air pressure to the supply ducts (see Figure 9.11). It is designed to be connected to the warm air heater outlet – and has a number of 'take-off' duct connections on the side, to allow connection of heating ducts. Where possible, the duct connections should not face the airflow from the heater outlet.

Warm air plenums used with down-flow heaters are designed to stand on the floor - and must be constructed to support the total weight of the heater (these are known as base or duct plenums).

Up-flow and horizontal (cross) flow plenums are similarly constructed, but are not weight bearing.

Attention: When you install open-flued warm air heaters, you MUST check that the joint between the heater and the plenum base is adequately sealed. Any discharge of air through inadequate sealing between the heater and plenum may adversely affect both combustion and chimney operation.

Figure 9.11 Warm air plenum

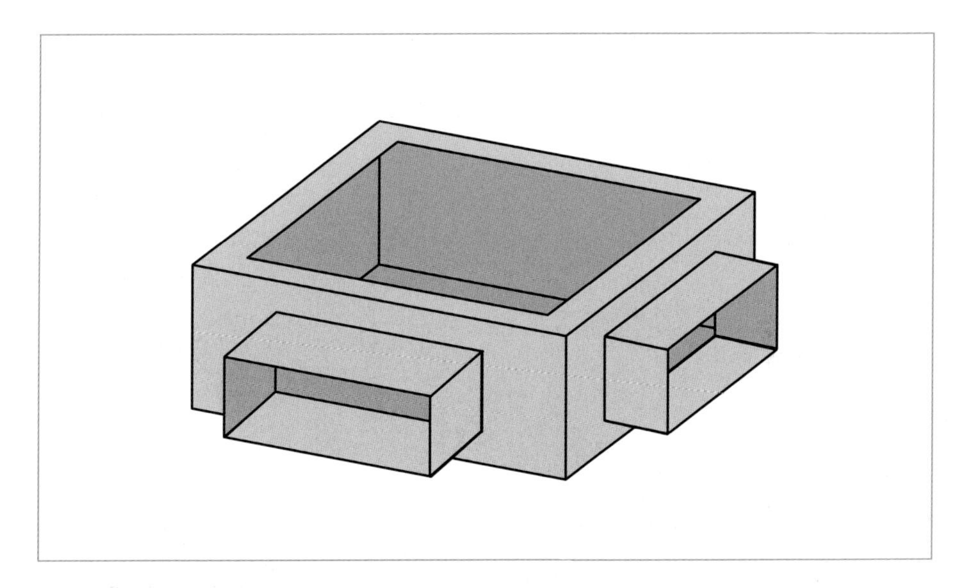

- make the same checks to both ductwork connections and blanked off outlets of the plenum to ensure they are adequately sealed. You must mechanically seal/secure all joints e.g. by self-tapping screws or by pop rivets (do not use duct tape for securing)
- if you fail to ensure this, it could result in the combustion becoming vitiated, creating poor combustion that results in CO being produced

It is also important that:

1. The heater is firmly supported; and
2. There is a 100% air path between the opening in the base of the heater and the plenum or base duct.

Balancing dampers – when to include them

Ideally, design warm air systems so that the main duct runs are in balance i.e. they have the same resistance to the airflow they are designed to handle.

- this degree of balance is not always possible (due to standard size ducts). So you may have to include 'balancing dampers' in the system design, to reduce the air velocity
- if, for example, a system has mostly long duct runs but includes one very short one, include a damper in the short run to balance the system
- as air velocity in rising ducts increases significantly, you may have to include dampers. You can position them in the ducts - or immediately behind registers or diffusers (see Figure 9.12)

Figure 9.12 Types of dampers

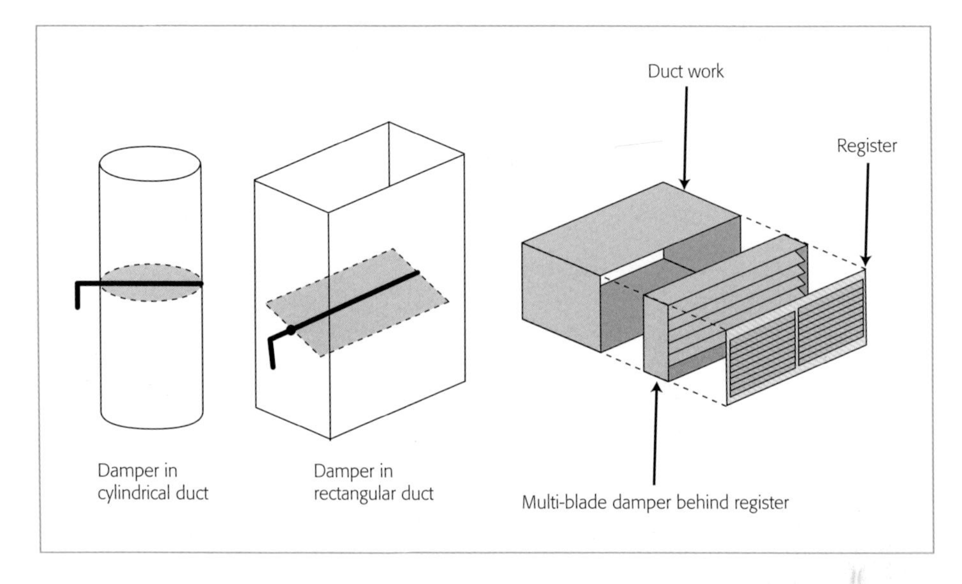

Types of duct systems – choose for location and needs

In most duct systems the cross-sectional area of a duct is kept roughly proportional to the volume of air passing through any point. So duct size should reduce as it branches. Normally you use one size for the main duct, with the branches reducing in size as necessary.

Air velocities in ducts must not exceed 4m/s. Typical ranges of duct size for which standard components and fittings are commonly used are shown in Table 9.2.

The type of duct system you install needs to meet the needs of a wide variety of residential installations, with your final choice depending on the building layout and economics. It may include more than one design to serve different floors/areas of a dwelling.

Stepped duct system

This is a main rectangular duct, with individual branches of either round or rectangular ducts. Figure 9.13 shows one main duct extending from the warm air plenum, but there could be two or more main ducts either connected to the warm air plenum - or branching from the main duct to form other main ducts.

- ensure take-off fittings from main ducts to individual areas of the dwelling slant in the direction of the airflow from the heater
- use reducing fittings where cross-sectional area changes are required

Table 9.2 Typical duct sizes

Cylindrical duct – Diameter (mm)	Rectangular duct (mm)	
100	150 x 100	200 x 200
125	150 x 200	200 x 300
150	150 x 250	200 x 400
175	150 x 300	200 x 500
	150 x 350	200 x 600

Figure 9.13 Stepped duct systems

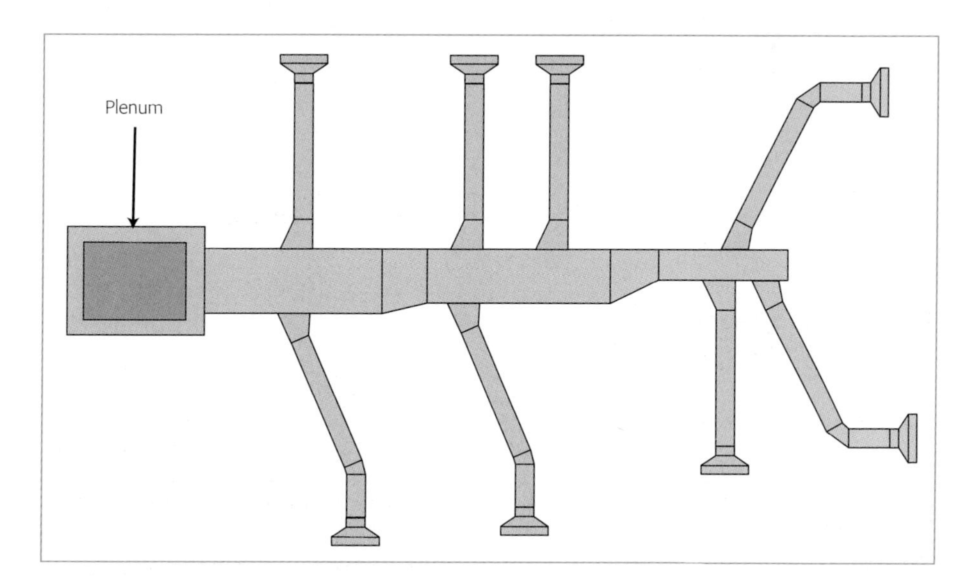

Extended plenum system

This is a simplified duct system where the main rectangular duct is the same size throughout its length, extending the warm air plenum (see Figures 9.14 and 9.15). Smaller round or rectangular ducts extend the system into individual areas of the dwelling.

This system can provide full perimeter heating and is best installed as part of the building structure.

The extended plenum system has two basic rules of conformity:

- main duct(s) should be no longer than 6m
- branch ducts should be no longer than 6m and contain no more than two bends (excluding boots)

Note: A boot is the transition from the round or rectangular duct to the diffuser/register (see Figure 9.16).

Figure 9.14 Extended plenum systems

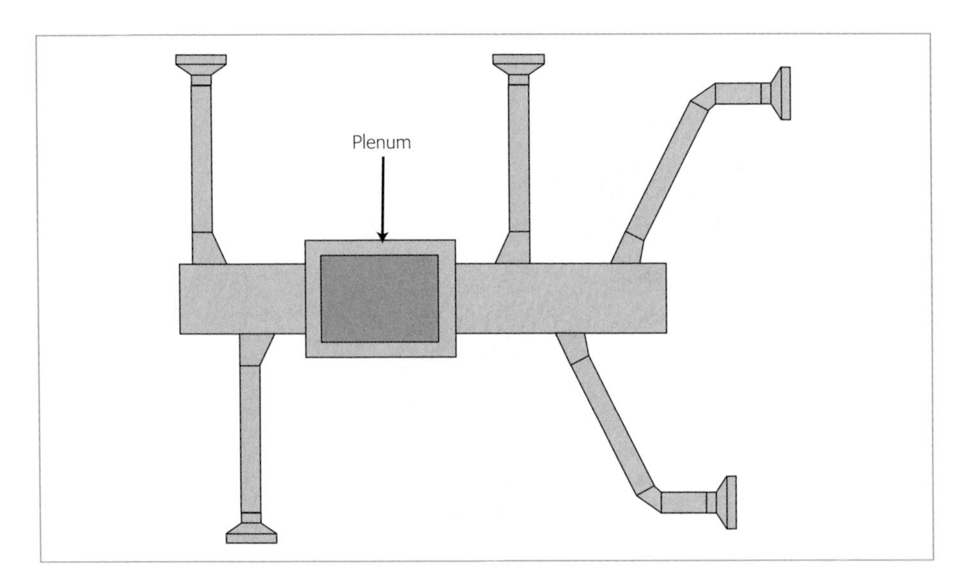

Figure 9.15 Extended plenum systems (perimeter)

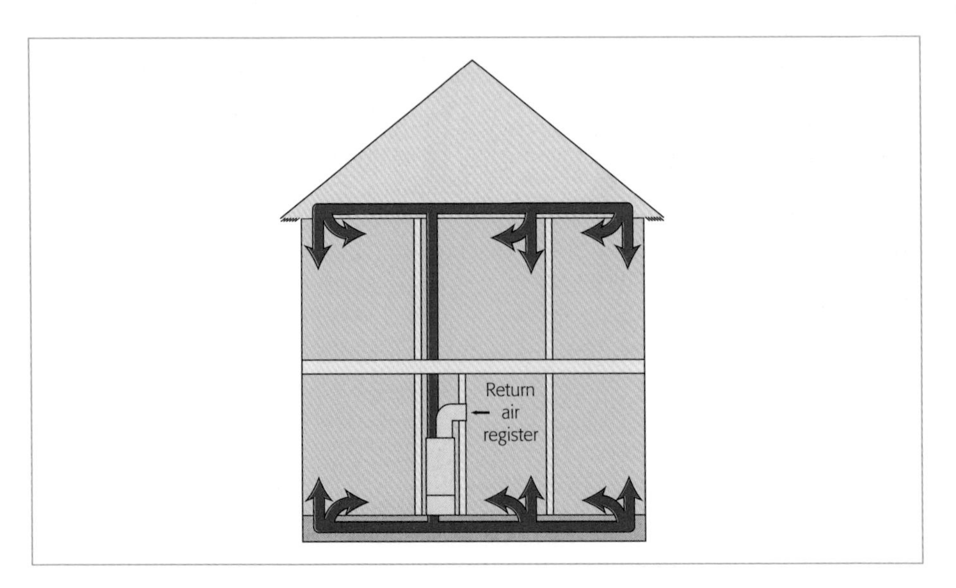

Figure 9.16 Boot fittings

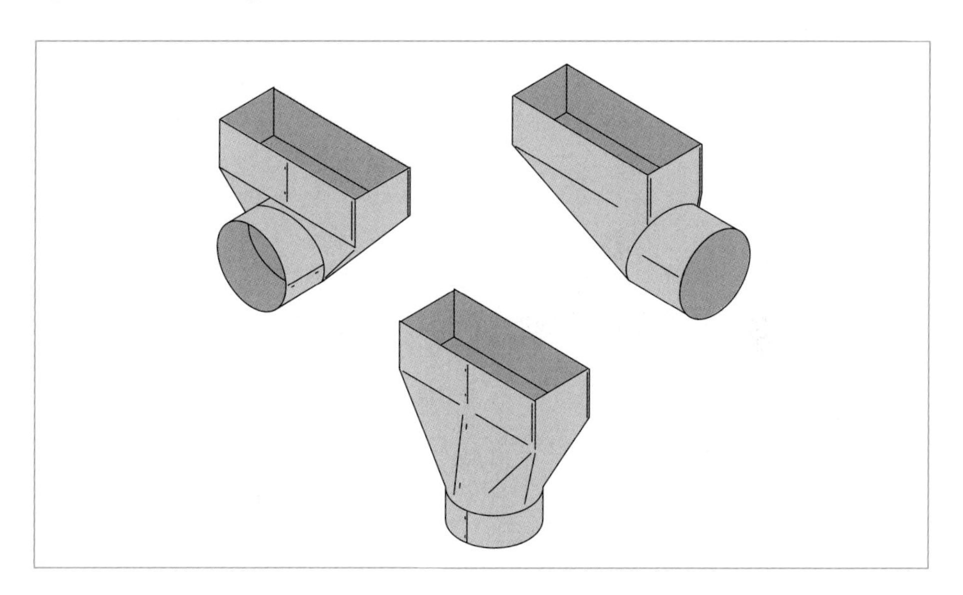

Radial (or stub duct) system

This system is also known as a Low Sidewall (stub duct) system. It can be less labour intensive as you can install it without under-floor ducts.

The warm air heater is centrally located with low sidewall registers to the living room, kitchen and hall. It can be extended to give similar heating to the first floor.

Cylindrical or rectangular ducts branch directly from the warm air plenum to individual areas of the dwelling.

Figures 9.17 and 9.18 are examples of their application.

This type of system has a basic rule of conformity:

- branch ducts should be no longer than 6m and contain no more than two bends (excluding boots)

Figure 9.17 Radial system

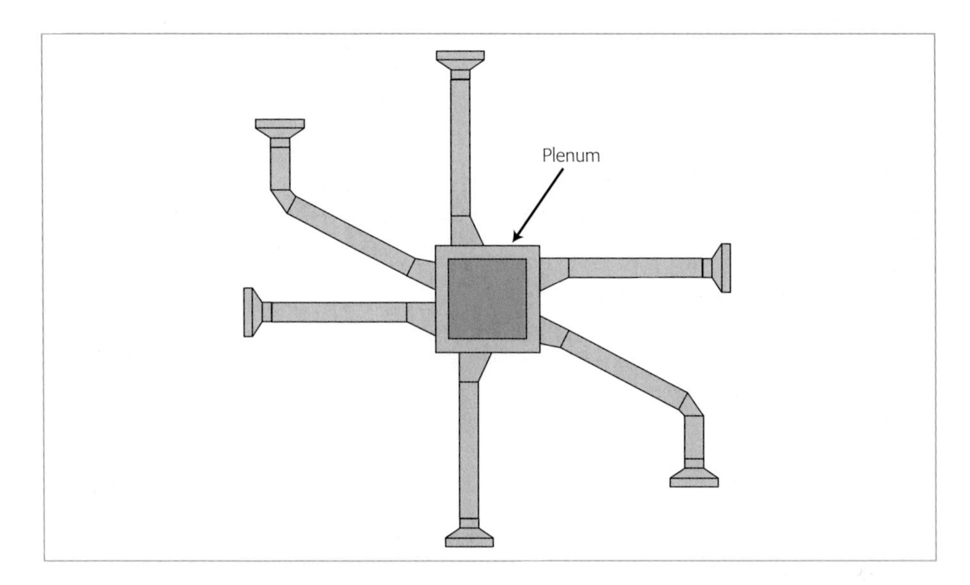

Figure 9.18 Low sidewall (stub duct) system

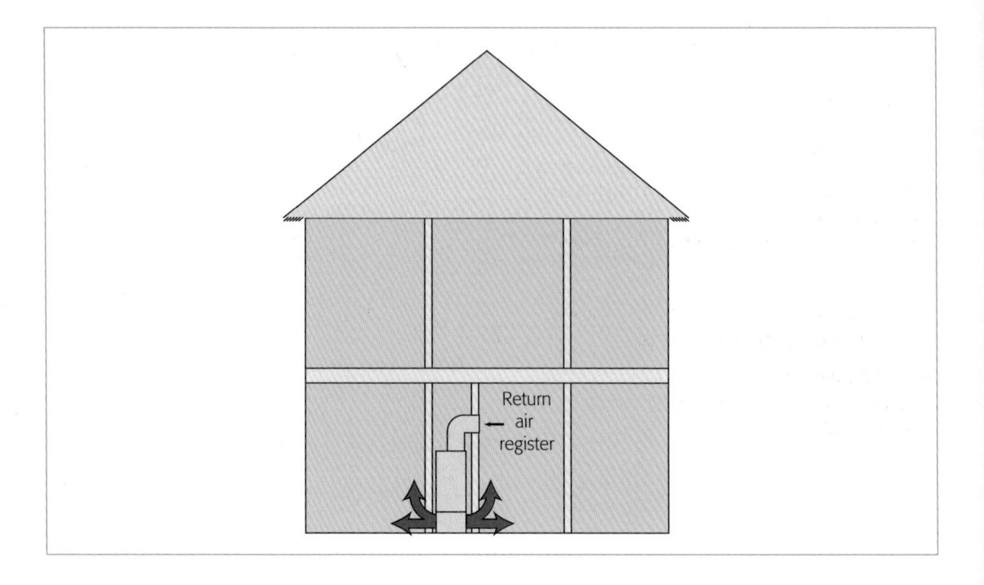

Why you need to balance the heating system

Balancing ensures heat is discharged in the various rooms/spaces according to their requirements. Systems that are out of balance may not give the required comfort conditions.

Some rooms/spaces may be too hot or too cold, whilst others are correctly heated. Balancing involves adjusting the flow of air and it can be carried out in both warm and cold weather.

A deflection type anemometer is recommended for recording the air velocity, as this automatically adjusts for changes in air density due to temperature changes.

An accurate thermometer capable of reading at least 70°C is also required.

Some problems to avoid

Some problems arise from duct lengths causing differing resistances, for example:

1. Very short duct runs, or registers fitted directly to the warm air plenum or base duct generally have a low resistance to air flow. But when installed together with some extended duct runs (which themselves have a high resistance) this creates an imbalance.

Note: You can overcome this by installing multi-blade dampers behind registers in short (low resistance) ducts to increase their resistance.

2. Too little resistance in ducts may cause overheating of upstairs rooms.

Note: You can overcome this by installing an adjustable damper in the rising duct(s) to increase resistance.

The balancing procedure

No two air measuring instruments are likely to give the same reading for a constant velocity. So it isn't possible to set velocities to specific design figures if readings can't be accepted as accurate by the measuring instrument used.

Consider design velocities as 'guide velocities'. Use this procedure to obtain a satisfactory system balance by setting velocities in correct proportions:

Table 9.3 shows the Air velocity factors.

Table 9.4 shows the Guide balancing velocities.

Table 9.5 shows the Final balancing velocities.

1. On warm air heaters with modulating controls, set the air circulation fan at maximum speed. Some warm air heaters with modulating controls (even temperature (ET) systems) have a switch labelled 'continuous' for this purpose. On basic models (those without an ET system) you need to bypass the fan switch, as it doesn't allow the fan to operate if no heat is detected. Carry this out following the manufacturer's instructions.
2. Carry out the measurements with the electrical supply ON, but the main burner OFF.
3. Partly close the balancing dampers of the diffuser/registers nearest to the warm air heater, and fully open those furthest away.
4. A worked example is shown in Tables 9.6 and 9.7 (see Note).

Note: For a genuine application, the heat required would be as given in the actual design and specification being installed. Air velocity factors in Table 9.4 are obtained from Table 9.3.

5. Establish the design balancing velocities from Table 9.4 and enter them in column 'A' of Table 9.5.
6. Measure the actual air velocity at each outlet and enter them in column 'B' of Table 9.5. Start at the diffuser/register nearest the heater and work away until the outlet on the longest duct is measured.
7. For each diffuser/register, divide the measured velocity by the design velocity (column 'B' divided by column 'A') and enter results in column 'C' of Table 9.5.
8. Find the average of column 'C' (add the results together and divide by the number of diffusers/registers) and note in column 'C' of Table 9.5.
9. Finally, multiply each design guide velocity in column 'A' by the average in column 'C' and enter results in column 'D' of Table 9.5.
10. Balance the system to the figures in column 'D' of Table 9.5 by adjusting the dampers and registers as necessary.

- following the balancing process, if the warm air heater has modulating controls (ET system) switch from 'continuous' to allow the fan speed control to operate - and complete commissioning as described in the manufacturer's instructions
- on basic models (those without an ET system), re-establish the fan switch and adjust the fan speed, to give a temperature rise across the heater which is close to the design value - and complete commissioning as described in the manufacturer's instructions
- if you measure a temperature rise across the heater, under or over 6°C of the design figure, adjust the fan speed to alter the temperature rise to the design figure

Table 9.3 Air velocity factors

Register size (mm x mm)	Air velocity factor
150 x 100	87
200 x 100	67
200 x 150	44
250 x 150	35
250 x 200	26
300 x 150	30
300 x 200	22
Diffuser size	**Air velocity factor**
57 x 250	93
57 x 300	78
57x 350	67
100 x 250	53
100 x 300	44

Note: The temperature rise across the heater is taken as the difference between the return air temperature (measured at the air filter) and the supply temperature measured as close as possible to the heater (i.e. at the nearest air diffuser/register to the heater).

- if necessary, alter the supply air temperature by adjusting the fan speed
- where diffusers/registers have dampers which have their travel distance limited by adjusting screws, adjust these to the maximum limit at the settings used
- finally, adjust the vanes on the diffusers/registers to give the desired direction of warm air flow. The system should stay ' in balance' unless others interfere with it

Table 9.4 Guide balancing velocities (figures shown are an example)

Room Column 1	Heat required (kW) Column 2	Register size (mm) Column 3	Air velocity factor Column 4	Guide velocity (column 2 x column 4) x 0.017 gives metres per second (m/s)
Lounge	2.5	250 x 150	35	1.49
Kitchen	1.3	200 x 100	67	1.48
Hall	1.5	200 x 150	44	1.12
Bathroom	0.9	200 x 100	67	1.03
Bedroom	1.9	200 x 150	44	1.42

Table 9.5 Final balancing velocities in metres per second (figures shown are an example)

Warm air outlet	Room	Guide design balancing velocities (m/s) Column A	Measured velocities (assumed) (m/s) Column B	Column B ÷ Column A Column C	Balancing velocities (m/s) Column A x average of Column C Column D
1	Lounge	1.49	1.4	0.94	(1.49 x 1.11) = 1.66
2	Kitchen	1.48	1.3	0.88	(1.48 X 1.11) = 1.65
3	Hall	1.12	1.6	1.43	(1.12 x 1.11) = 1.24
4	Bathroom	1.03	1.3	1.26	(1.03 x 1.11) = 1.14
5	Bedroom	1.42	1.5	1.06	(1.42 x 1.11) = 1.58
				Total = 5.57 Average = 5.57 ÷ 5 = 1.11	

Warm air heater replacement – 10

10 – Warm air heater replacement

Figures

What you need to know

- when you plan to replace a warm air heater, the new installation must conform to the current Gas Safety (Installation and Use) Regulations, the appliance manufacturer's instructions and current British standards
- also refer to the installation guidelines of **Part 9 – Warm air heating** of this manual
- this guidance generally applies to replacing open-flued gas-fired warm air heaters in domestic premises of one or two storeys
- buildings of more than two storeys have additional requirements (see **Part 9 – Warm air heating – Types of buildings**)

For commissioning, servicing, maintenance and fault finding procedures see **Part 11 – General installation details – Warm air systems**.

Warm air heating systems were at their most popular in the 1960s and 1970s and were installed to the British Standards then in force. Many of these installations will be due for replacement and upgrading to current installation standards.

- the first point for you to consider when you replace a warm air heating installation is whether the design of the existing system is satisfactory

  you can do this by asking the user if the existing system provides the required amount of heat to the rooms or internal spaces (hall, landing etc). Depending on what the user says, you may need to totally re-design the heating system - or simply balance it correctly to meet their needs, once you have installed the new heater

Some older types of warm air heaters are controlled by a room thermostat, which operates the main burner and circulation fan as soon as heat is called for, causing cold air to be delivered until the heat exchanger is up to temperature.

When satisfied, the room thermostat shuts down the burner and circulation fan together, leaving a residue of heat in the heater.

When introduced, the fan delay switch greatly improved comfort levels: by delaying the circulation fan operating until heat was available. Also the circulation fan runs on after the burner has switched off, distributing all residual heat within the heater.

Even Temperature (ET) systems were then introduced to modulate both the burner output and circulation fan speed - by constantly sensing heat requirements.

How to overcome some problems with existing systems?

Some quite simple methods are to carry out a service on the appliance (see **Part 11 – General installation details – Warm air systems – How to carry out servicing**) or you may need to replace a very old appliance and re-use the existing ductwork system, in which case consider the following points:

1. Warm air registers may have been fitted incorrectly so that the closeable vanes are directing warm air upwards instead of downwards. The user may have caused this by removing the register to decorate - and re-fitting it the wrong way round.
2. Balancing dampers (see **Part 9 – Warm air heating – Duct systems – points to consider – Balancing dampers – when to include them**) may not have been fitted at the time of installation to allow the system to be correctly balanced. You quite often find house bricks placed in ducts (by the original commissioning engineer or builder installing the ducts) for balancing.
3. There may be draughts due to air velocity at registers being too high.
4. Ducts and registers may have been undersized and therefore unable to deliver sufficient heat.

5. Some parts of the room/space may be too cold due to badly positioned warm air outlets causing poor warm air circulation.

How to choose the right replacement heater

Before you can do this, you need to assess the existing system and know if the new system is to be extended to other unheated parts of the dwelling, take into account the following points:

1. Heat requirements – make heat loss calculations for each room/space to be heated - to determine the heater size (see **Part 9 – Warm air heating – Correct design and installation of warm air systems essential**).
2. Duct system – inspect the existing duct system to ensure it's in good condition and sized correctly for the heat requirements (see **Part 9 – Warm air heating – Duct systems – points to consider**).
3. Registers/diffusers must be correctly sized for the heat requirements and operate satisfactorily.

If you use the existing system, deciding which replacement heater to install depends largely upon the existing position and condition of the following major components of the installation:

- warm air plenum
- type of air flow (upflow, downflow or horizontal (cross) flow)
- type of flue connection on existing heater e.g. front or rear
- chimney system
- return air arrangement
- water heating requirements

Warm air plenum: what you need to do

- the prime objective when replacing a warm air heater is to ensure any leakage of air from the existing plenum and ductwork system is kept to a minimum. This is especially important if you fit the heater in a compartment
- it is essential that the circulation fan of the heater is prevented from interfering with the operation of the burner and the chimney system
- thoroughly examine all existing ductwork joints - and where necessary make them mechanically sound i.e. by using self-tapping screws/pop rivets and finally sealing them with duct tape
- the existing plenum should be in good, sound condition if you are to re-use it; otherwise replace it

The warm air plenum is a 'box' designed to equalise air pressure and temperature before it's distributed into the dwelling through a network of ducts connected to it.

- you will need to identify the heater type, either upflow, downflow or horizontal (cross) flow (see **Part 9 – Warm air heating – Types and application of heaters – choose to suit dwellings**)
- the heater type determines the position and construction of the plenum. Warm air plenums used with downflow type heaters are designed to allow the heater to stand directly on them - and must be constructed to support its weight
- upflow and horizontal (cross) flow heater plenums are similarly constructed, but are not weight-bearing

Figure 10.1 Fitting new heater to existing plenum

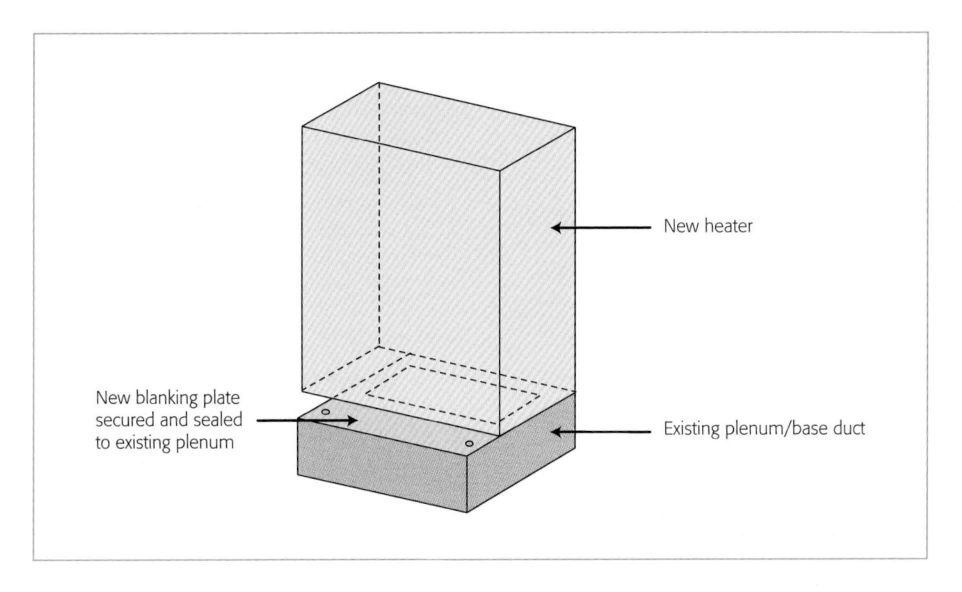

Modern warm air heaters differ in dimensions to older types and are generally smaller.

- if you re-use the existing plenum, you may need to adapt it to fit the new heater and to blank off openings left by the old heater (see Figure 10.1). It is recommended you use the warm air heater manufacturer's adapter kit, which contains specific instructions on how to do this
- secure the blanking plate mechanically to the plenum i.e. with self-tapping screws or pop rivets, and seal it with a self adhesive foam strip, Neoprene sealing strip, or a suitable sealing compound (see Note)

Note: Duct tape is for sealing - not for securing.

When you install open-flued warm air heaters, any discharge of air through inadequate sealing between the new heater and the existing plenum may adversely affect combustion and chimney performance.

It is therefore extremely important that you make an effective seal between the heater and the plenum, you must also ensure:

1. The new heater is firmly supported; and
2. There is a 100% air path between the opening in the base of the new heater and the existing plenum or base duct.

Generally three types of air flow

Fan-assisted heaters are categorised with reference to the air flow movement through them and are generally divided into three types (see **Part 9 – Warm air heating** – Figure 9.2):

1. Upflow – generally free-standing, taking air from low level through a heat exchanger, to discharge to high level ducting.
2. Downflow (the most commonly used) - generally free-standing, taking air from high level through a heat exchanger to discharge to low level ducting.
3. Horizontal (cross) flow – generally wall-mounted, taking air through a heat exchanger to discharge to side ducting.

Note: If the replacement heater has a different type of airflow than the existing one, you may have to carry out major alterations to the duct system.

Type of open-flued chimney connection

- once installed, correctly flue the replacement heater. Heaters are designed with chimney connections located either at the back, top rear or top front (see **Part 9 – Warm air heating** – Figure 9.1)
- if the replacement heater does not have a similar chimney connection, you need to make alterations which may affect the chimney performance
- avoid excessive use of chimney bends - so the new heater's chimney connection must be similar to the existing one

Chimney system – open-flued

General

As warm air heating system efficiency relies on short ductwork runs, most warm air heaters are positioned in the centre of the dwelling.

A typical existing chimney system generally runs from the heater up through a bedroom and into the loft area, terminating to atmosphere with a ridge terminal. Or it may pass directly through the pitched/flat roof terminating with an approved terminal.

When you replace the old heater with a new one, pay particular attention to the chimney system as it may involve:

1. Connecting to the existing chimney system; or
2. Completely renewing the chimney system.

Whichever option you select, the chimney system must conform to the warm air heater manufacturer's installation instructions (see also the current Essential Gas Safety – Domestic – Part 13 for further guidance).

Connecting to an existing asbestos cement chimney system

Many early open-flued warm air heaters were flued using asbestos cement flue pipe and fittings (see **Warning** on page 172).

If you come across this material, you can re-use this type of chimney system providing you carry out these basic checks beforehand:

1. Only one appliance must be connected to the chimney system.

Figure 10.2 Ridge terminal adapter ('R' type adapter)

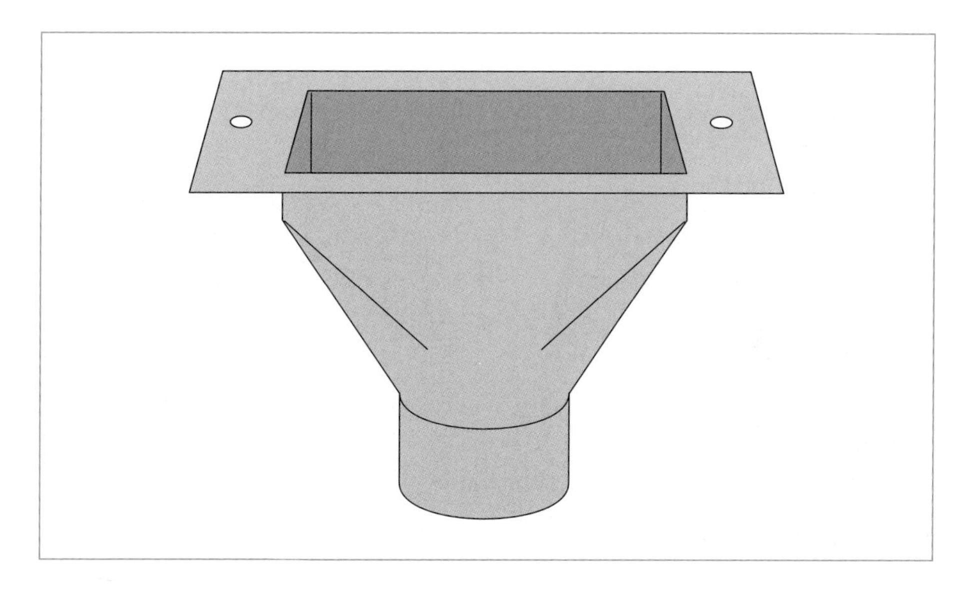

2. Check the entire route of the chimney to ensure that:
   a) There are no bends in the chimney greater than 45 degrees to the vertical which could adversely affect the performance of the chimney or result in a blockage.
   b) You inspect all joints - and make good where necessary.
   c) The chimney is correctly supported.
   d) The chimney is in good condition for re-use.
3. Examine the termination and replace as necessary. Older ridge terminal adapters ('R' type adapters) which provide the connection between flue pipe and ridge terminal, (see Figure 10.2) are known to have securing bolts that are subject to corrosion.

   Check their security/suitability to ensure they are fit for further use. In the past, some securing bolts were manufactured using plastic materials. Replace these too.

Figure 10.3 Typical low resistance ridge terminations

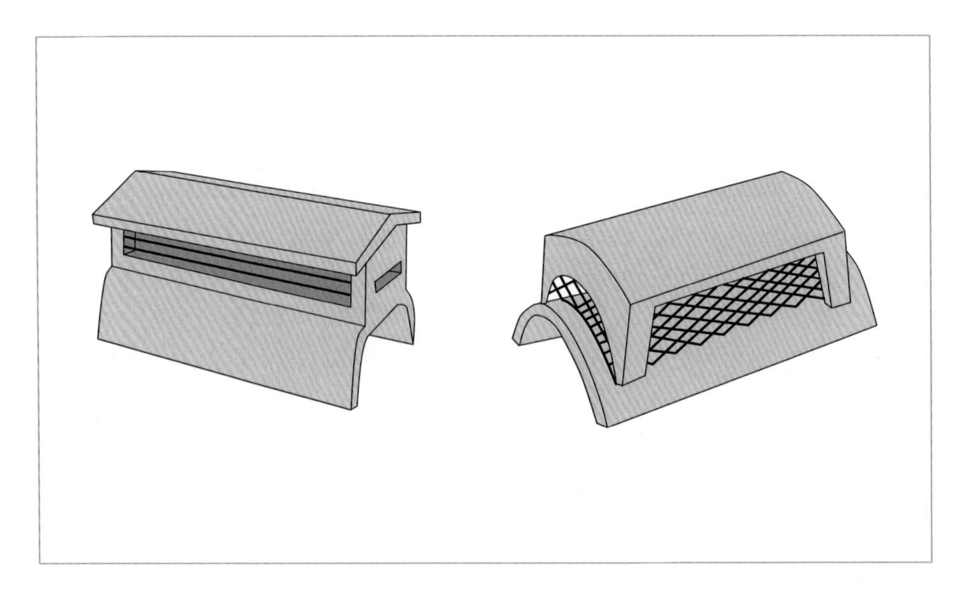

4. Where an old type ridge terminal is fitted, these are known to have a high flow resistance. Consider fitting a new, less restrictive type (see Figure 10.3).

   High flow resistance ridge terminals are typically open on two sides (front and back) and generally follow the profile of the ridge tiles. Low flow resistance ridge terminals are generally open on all four sides and project above the profile of the ridge tiles.

Upgrade existing chimney systems that don't comply with current standards, or are in poor condition. You generally do this by connecting double-walled metallic flue pipe/fittings to the 'sound' part of the existing asbestos cement chimney system.

Special fittings are normally available from chimney system manufacturers to connect to the existing asbestos cement chimney. This method generally uses a socket type fitting that requires a flue-jointing compound to be applied to the connection of the existing chimney.

- note that if you use excessive flue jointing compound, this may force some of the compound out of the joint and into the flue pipe – this could obstruct and so reduce the internal diameter of the flue

Warning about asbestos removal:

- **asbestos removal is specialist work which must only be carried out by approved businesses as defined in the Control of Asbestos at Work Regulations**
- **in accordance with the Health and Safety at Work etc Act, apply special safety precautions when you work on an existing asbestos cement flue pipe/fitting**
- **for further detailed advice, contact the Health and Safety Executive Information line on 0845 345 0055. Alternatively, visit www.hse.gov.uk/asbestos**

Note: When you connect to an existing chimney or part-renew the chimney system, removing the existing warm air heater will cause some movement, however small, of the existing chimney system.

- when you complete the installation of the replacement heater, carry out items 2 b) c) d) and 3 of **Connecting to an existing asbestos cement chimney system** again

Connecting to an existing lined chimney system

- if you connect to a flexible metallic liner, replace the existing liner – unless you consider it will operate safely throughout the life span of the new warm air heater
- under normal operating conditions, a liner complying with BS EN 1856-2 should operate safely for at least the operational life span of an appliance (normally 10-15 years). So if the existing warm air heater is older than 10 years, it's recommended you replace the liner

Connecting to an existing twin walled chimney system

If you connect to an existing twin walled chimney system and the new and existing chimney systems are by different manufacturers, then use an approved adapter to connect the two chimney systems together.

When to completely renew the chimney system

If you have any doubts regarding the existing chimney system's operation or compliance with current standards, or if you suspect its poor condition, renew the complete chimney system (see the current Essential Gas Safety – Domestic – Part 13 for further guidance)

Chimney system – room-sealed

Free-standing/wall mounted

- replacing free-standing or wall-mounted room-sealed warm air heaters can sometimes be a simple operation, especially if you replace an appliance in the same position. You may only need to make good the building structure internally and externally to suit the new terminal assembly
- it is important you site the terminal correctly (see **Part 11 – General installation details – Warm air systems – Natural draught room-sealed warm air heaters** and **Fanned draught room-sealed warm air heaters**). See also the current Essential Gas Safety – Domestic – Part 13 for further guidance
- alternative locations may necessitate major alterations, especially to the warm air ductwork system and return air path
- when only slight alterations are necessary, a room-sealed appliance may be more suitable. For example, you can sometimes replace an open-flued wall mounted warm air heater fitted under stairs, with a room-sealed down flow model. You may need to adapt the warm air plenum to suit the new heater (see **Warm air plenum: what you need to do** in this Part)

Se-duct/U-duct

When you replace Se-duct and U-duct warm air heaters, the new spigots may not match up with the existing holes.

To reduce installation time and to avoid damage to the ducts by re-cutting, warm air heater manufacturers normally supply a special transfer box manufactured from a corrosion resistant metallic material for a variety of old warm air heater models, complete with detailed installation instructions. These transfer boxes may be offset or right-angled depending on the model (see Figure 10.4) and are supplied in pairs (one each for flue connection and air entry spigots).

Note: Make good any damage to the building structure.

For more information on Se-duct and U-duct chimney systems, see the current Essential Gas Safety – Domestic – Part 13.

Chimneys in voids

Chimneys – room-sealed and open-flued – run within voids (ceilings, floors, walls, etc.) have become a concern to the industry, where these systems cannot be inspected visually on a periodic basis as required by the GSIUR.

As a result, Gas Safe Register has produced a TB, which provides guidance on the issues and what steps are required to be taken to ensure continued safety – TB 008 (Edition 2.1) 'Room-sealed fanned-draught chimney/flue systems concealed within voids'.

Gas installers/businesses are urged to obtain a copy of the TB – freely available to registered businesses via Gas Safe Registers web site – and ensure they read and implement its requirements immediately (see **Part 8 – General installation details – Wet central heating – Chimneys in voids** – for further guidance.

Replacement of natural convection (brick central) models

Type-for-type replacement

You can replace existing gas, oil fired or electric natural convection 'brick central' model warm air heaters using the modern equivalent gas-fired appliance.

Fan-assisted type replacement – general guidance

- if a fan-assisted gas-fired model is specified as a replacement for a natural convection 'brick central' model warm air heater, follow the particular appliance manufacturer's instructions
- general guidance on this subject is as follows (see also Figure 10.5):

Gas-fired model

1. Dimensions – make sure the replacement heater will pass through the compartment door. If not, check you can structurally alter the door/frame.
2. Compartment – you may have to upgrade to meet current requirements/standards (see **Part 9 – Warm air heating – Compartment installations – what you must do**).
3. Heater Output – where required, make allowance for heating additional rooms, for example, bedrooms.
4. Chimney – install a suitable chimney system by either connecting to the existing chimney (if suitable), part renewing or completely renewing (see **Chimney system – open-flued** in this Part). Due to the design of natural convection installations (see **Part 9 – Warm air heating – How it operates**) the siting will generally be unsuitable for a room-sealed heater.

Figure 10.4 Transfer boxes

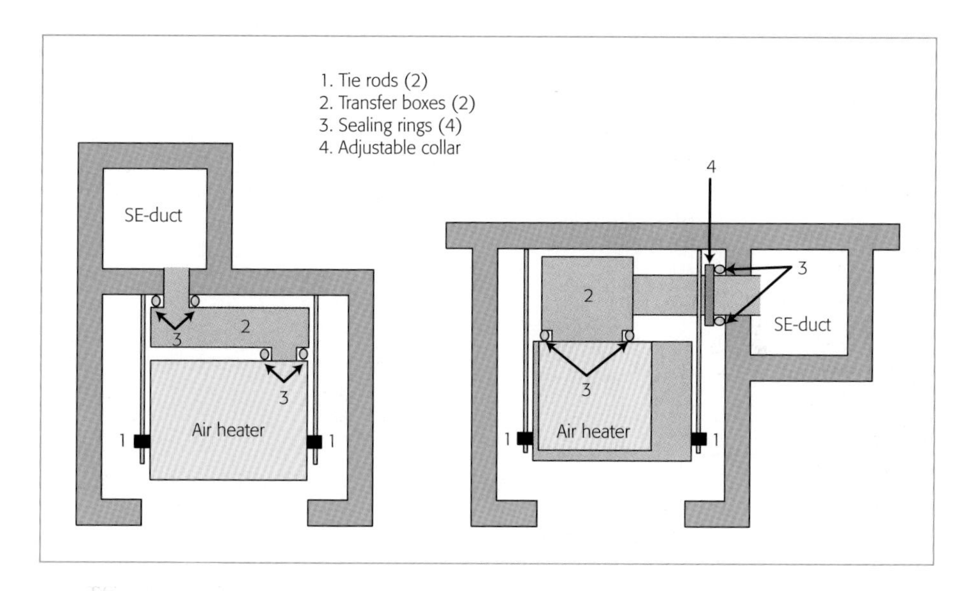

5. Return air – with a natural convection 'brick central' model warm air heater, this type of installation is not normally fitted with a return air system. So provide a positive return air duct and possibly a return air path (see **Part 9 – Warm air heating – Return air (re-circulation)**). Consider using an existing high level warm air convection grille for the return air duct connection to the new heater.
6. Compartment/combustion ventilation – check it's in accordance with the manufacturer's instructions.
   See also the current British Standard for ventilation requirements: BS 5440-2 for further guidance.
7. Warm air outlets – consider using the existing low-level return air openings.
8. Plenum (or base duct) – provide this for the new heater and install ductwork to include upper floors if the construction of dwelling will permit this.
9. Electrical supply – generally natural convection warm air heaters (other than electrical models) do not have electrical components. Check availability (see **Part 11 – General installation details – Warm air systems – Electrical connections**).

Figure 10.5 Replacement of a gas, oil or electric brick central heater with a fan-assisted model

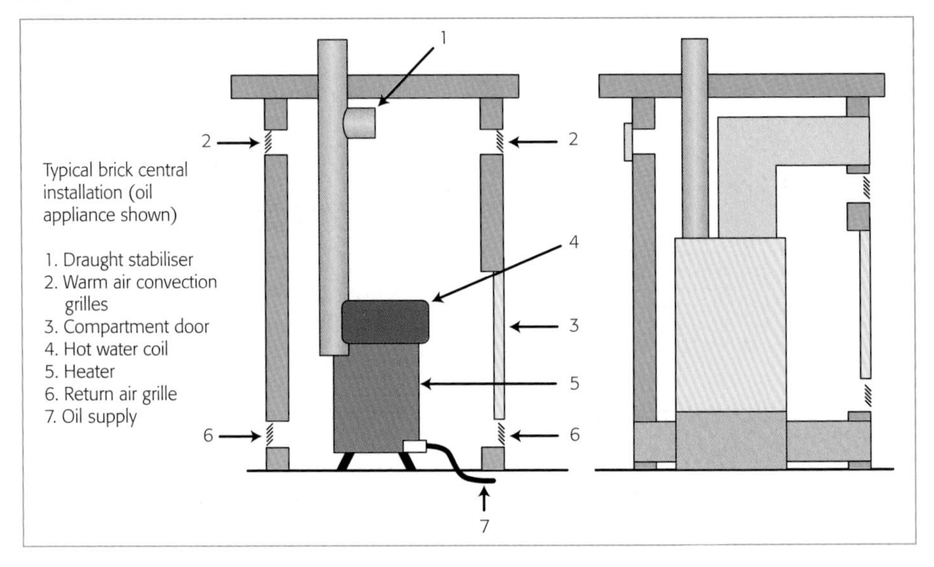

Oil-fired model

You can replace existing oil-fired natural convection ('brick central') systems following the gas-fired guidance above. However, the existing chimney system requires additional attention as follows:

1. To be acceptable, the chimney system must conform to the requirements of BS 5440-1.
2. If you know the existing chimney performed unsatisfactorily with a previous appliance or fuel, sweep and examine it. Correct any faults found.
3. Remove any register plates, restrictor plates or dampers or permanently secure them in the fully open position, to leave the main part of the flueway unobstructed.
4. Where necessary, line the existing chimney using a BS EN 1856-2 approved flexible metallic liner.
5. Fit a suitable flue terminal.

Note: If you have any doubts about the operation or condition of the existing chimney system, it's recommended you completely renew it.

Electric model

You can replace existing electric natural convection ('brick central') systems using the gas-fired guidance above.

However you need to consider these additional requirements:

1. Gas supply – ensure availability.
2. Chimney – install a suitable chimney system. Depending on the location, it may be possible to install a room-sealed warm air heater model.

Figure 10.6 Replacement of fan-assisted electric warm air heater

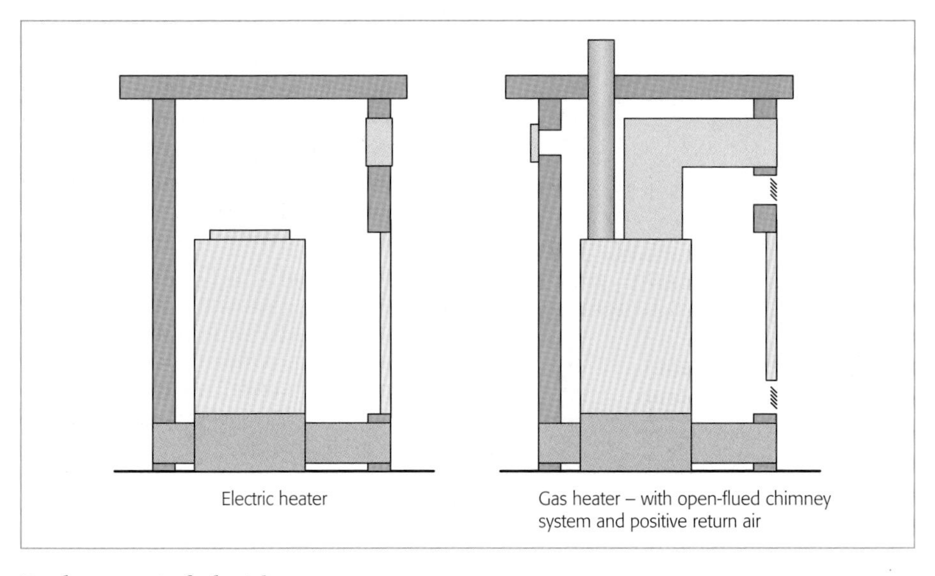

Replacement of electric fan-assisted models

If a fan-assisted gas-fired model is specified as a replacement for an existing fan assisted electric warm air heater fitted in a compartment (see Figure 10.6), then follow the particular appliance manufacturer's instructions, but general guidance on this subject is as follows:

1. Gas supply – ensure its availability.
2. Dimensions – make sure the replacement heater will pass through the compartment door. If not, check that you can structurally alter the door/frame.
3. Compartment – you may have to upgrade to meet current requirements/standards (see **Part 9 – Warm air heating – Compartment installations – what you must do**).
4. Heater output – where required, make allowance for heating additional rooms e.g. bedrooms.
5. Chimney – install a suitable chimney system (see **Chimney system – open-flued** – in this Part).
6. Return air – existing electric fan-assisted heaters may not have a return air system. So you need to provide a positive return air duct – and possibly a return air path (see **Part 9 – Warm air heating – Return air (re-circulation)**).
7. Compartment/combustion ventilation – check it is in accordance with the manufacturer's instructions. See also the current British Standard for ventilation requirements: BS 5440-2 for further guidance.
8. Warm air outlets – consider using the existing low-level return air openings.

9. Plenum (or base duct) – you need to adapt the existing plenum to suit the new heater – or provide a new one (see **Warm air plenum: what you need to do** in this Part).
10. Ductwork – may be limited to existing rooms/spaces surrounding the heater. Consider extending the ductwork system to upper floors for example, if the construction of dwelling permits this.
11. Electricity – check suitability (see **Part 11 – General installation details – Warm air systems – Electrical connections**).

Return air arrangement – design to avoid problems

- on an existing open-flued warm air heating system, an incorrectly designed or installed return air duct/system may create a negative pressure within the heater compartment and draw POC down the chimney, through the heater and into the property via the ductwork system, with dire consequences (see **Attention**)
- similar problems can occur when the system is not provided with a positive return air connection or an adequate return air path back to the heater
- provide a full and unobstructed return air path to the heater from all heated rooms and spaces (except kitchens, bathrooms and toilets). This allows air from heated rooms/spaces to be returned to the heater for re-heating and re-circulation
- you provide the return air path itself through return air grilles connected by ducts to the heater. Relief air openings allow air to move to the return air grilles from other rooms/spaces – but not directly between bedrooms

Sizing, positions and installation are dealt with in **Part 9 – Warm air heating – Return air (re-circulation)**.

Attention: The return air duct/grille arrangement will have been designed for the existing heater.

- **so to ensure the safe operation of the new appliance, check the return air arrangement; and**
- **where necessary upgrade to deal with the requirements of the new appliance in accordance with manufacturer's instructions – failure to do so could result in an unsafe situation**

Balance the existing system

Balancing is carried out to ensure that the warm air is discharged to the various rooms/spaces to match their requirements.

- carry out balancing in accordance with the manufacturer's installation instructions
- refer to the general guidelines in **Part 9 – Warm air heating – Why you need to balance the heating system**

General installation details – Warm air systems – 11

11 – General installation details – Warm air systems

Figures

Tables

Introduction

This guidance is general to the installation of new and replacement warm air heating appliances and applies to fan-assisted models (which are the most commonly used type). However, electrical connections, chimneys, gas supply and ventilation may also be common to natural convection (brick central) models.

Attention: For particular guidance on the installation, commissioning, servicing, and fault finding of circulators see the current Gas Installer Manual Series – Domestic – 'Water heaters'.

Warm air heater locations

To avoid repetition, permitted locations for warm air heaters, either open-flued or room-sealed is as described in previous Parts of this manual for boilers – refer to **Part 2 – Open-flued boilers** and/or **Part 3 – Room-sealed boilers – Restricted locations – for safety reasons**, as appropriate.

For other locations including compartments, understairs, roof spaces, etc. refer to **Part 9 – Warm air heating – Installation – what you need to consider.**

For warm air heaters located in rooms containing a fixed bath or shower refer to **Part 8 – General installation details – Wet central heating – Gas appliances in bathrooms – electrical zoning requirements.**

With reference to electrical issues, including safe isolation practices, refer to **Part 8 – General installation details – Wet central heating – Electrical connections – you are responsible for compliance** and **Electrical isolation – essential for safety.**

Commissioning: your responsibility

You are responsible for ensuring that all work is carried out in accordance with the relevant Regulations, the gas appliance and installation operate in a safe and satisfactory manner and:

- when you connect a gas supply to an appliance, the Gas Safety (Installation and Use) Regulations require you to commission that appliance
- unless you can complete this work immediately, you must disconnect the appliance from the gas supply and label it accordingly
- you must test all gas fittings forming part of the installation for gas tightness - and purge of air (see the current Essential Gas Safety – Domestic – Parts 6 and 15 for further guidance)

You will find additional information on LPG tightness testing in the current Gas Installer Manual Series – Domestic – 'LPG – Including Permanent Dwellings, Leisure Accommodation Vehicles, Residential Park Homes and Boats'.

- you need to examine the warm air heater installation to determine whether a sufficient positive return air path exists from all heated rooms back to the warm air heater 'collection area'. This area will often be the hall or landing of the dwelling
- on open-flued appliances, a sealed positive return air duct must connect the return air grille(s) to the return air connection of the heater
- you must follow the manufacturer's commissioning instructions supplied with the warm air heater and (where fitted) circulator
- you may use the general procedure that follows for appliances covered in this Part

- where a circulator is an integral part of the appliance, commission in accordance with the manufacturer's instructions

You will also find guidance in the current Gas Installer Manual Series – Domestic – 'Water Heaters'.

General procedure

1. Check the ventilation requirements are correct and in accordance with the manufacturer's instructions (see **Ventilation – when and why needed** in this Part. See also the current British Standard for ventilation requirements: BS 5440-2 for further guidance).
2. Check the chimney termination is correct (see the current Essential Gas Safety – Domestic – Part 13 for further guidance).
3. Where applicable, carry out a flue flow test (see the current Essential Gas Safety – Domestic – Part 14 for further guidance).
4. Check electrical connections (see **Part 8 – General installation details – Wet central heating – Electrical connections – you are responsible for compliance**).
5. Test all appliance gas connections with non-corrosive leak detection fluid (LDF).
6. Check the air filter, air circulation fan and the fan compartment are free of obstructions.
7. Check the return air and air relief openings are correctly sized and unobstructed.
8. Light the warm air heater in accordance with manufacturer's instructions.
9. Check that the operating pressures, gas rates or both are in accordance with the appliance data plate. Adjust as necessary (see the current Essential Gas Safety – Domestic – Part 11 for further guidance).
10. Check all the burners cross light - and the flame picture is satisfactory in terms of stability, structure and colour.
11. Check for correct operation of all control valves and that the ignition system (if applicable) operates correctly.
12. If necessary, adjust the pilot flame, to envelop the thermocouple tip. Ensure that it maintains the flame supervision device (FSD) correctly. If the pilot light is extinguished, do not try to re-light it for 3 minutes.
13. Check the 'fail safe' of the FSD in accordance with the manufacturer's instructions (see the current Essential Gas Safety – Domestic – Part 12 for further guidance).
14. Following manufacturer's instructions, check that the high limit (or overheat) switch operates correctly by operating the heater with the main burner alight and the fan disconnected. The heater should shut down within 3 to 5 minutes.
15. Carry out a spillage test (see **Spillage testing** in this Part and also the current Essential Gas Safety – Domestic – Part 14 for further guidance).
16. Check the warm air heater thermostats and air temperature thermostat are operating satisfactorily. Depending on the type fitted, turn to the lowest setting, or 'off' position. The warm air heater should go off.
17. Ensure all diffusers/register shutters open and close freely.
18. Check the warm air heater operation for undue noise arising from mechanical defects or faulty installation such as loose electrical connections, and loose motor or fan mountings.
19. Ensure that any compartment warning labels are correctly fixed (see the current Essential Gas Safety – Domestic – Part 10 for further guidance).

20. Instruct the user on how to operate the appliance and controls.
21. Where appropriate, instruct the user on the need to clean the air filter regularly.
22. Complete all relevant documentation – Benchmark for example – and leave all instructions with the user.
23. Advise the user that the warm air heater will require servicing/safety checks at a minimum of 12 monthly intervals or at intervals specified in the manufacturer's instructions.
24. Advise the user of any appliance/installation defects in writing. If necessary, follow the current Gas Industry Unsafe Situations Procedure (see also the current Essential Gas Safety – Domestic – Parts 8 and 10 for further guidance).

Note: A gas appliance in normal use will require servicing and safety checking at 12 monthly intervals. This period depends on the amount of use and the type of room or space it is installed in. It may, therefore, require servicing at intervals less than 12 months.

Spillage testing

- carry out a spillage test(s) in accordance with the manufacturer's instructions to ensure that the warm air heater chimney system is operating safely

  The test(s) should also prove that suction (negative pressure) created by the air circulation fan on the return air arrangement, does not create a sub-atmospheric pressure in the room/compartment, that can cause spillage of the POC, from the warm air heater and/or circulator (where fitted).
- also make checks on any other open-flued appliance (gas, oil or solid fuel) installed in the same or adjoining rooms that may also be affected by the operation of the warm air heater

During the test(s), consider the operation of other fans or similar extract devices that are present in the same or adjoining rooms within the dwelling, e.g. cooker hoods, tumble dryers, bathroom/toilet fans, etc.

- carry out the appliance manufacturer's spillage test with the appliance on and under full operating conditions i.e. outer case on and/or any door to the room or compartment where the warm air heater is installed, closed
- during the test you should be able to see whether the test is satisfactory or unsatisfactory. However, on some installations the compartment may be so confined that you are unable to carry out the test with the compartment door closed
- if this is so, consult the appliance manufacturer for a suitable test procedure applicable to their appliance
- if there is no such procedure, carry out the spillage test on a 'best endeavour' basis, which may require you to carry out the test with the compartment door open

Where the appliance manufacturer gives no particular guidance, follow the following advice:

If the draught diverter is accessible, carry out a spillage test following the procedure given in the current Essential Gas Safety – Domestic – Part 14, if the draught diverter is not accessible – (e.g. 'slot-fix'/storey height heaters):

1. Pre-heat the appliance to normal operating temperature.
2. Extinguish the main and pilot burners (ensure the air circulation fan is running while you carry out the spillage test; where necessary use the 'summer' fan control to ensure the fan is operating).
3. Using an ignited smoke pellet on a non-combustible support, introduce smoke into the combustion chamber area of the heat exchanger (see Note overleaf).

4. Check there is no spillage evident by looking at the general area of the down draught diverter on the warm air heater.
5. If spillage is evident, carry out further investigation and remedial work before you re-test the appliance.

If you have any doubt regarding the safe use of the appliance, follow the current Gas Industry Unsafe Situations Procedure
(see also the current Essential Gas Safety – Domestic – Parts 8 and 10 for further guidance)

Note: Do ensure the size of the smoke pellet you use is not too great in relation to the amount of smoke it produces. In the case of an open-flued appliance, the smoke produced may be greater in volume than the heat exchanger and chimney of the appliance can handle.

This can cause smoke to 'spill' from the burner or down draught diverter positions giving a false result. Once the smoke has 'spilt' into the area surrounding the appliance it is difficult for you to detect whether or not there was any spillage in the first place.

Flue testing with decorative recirculatory ceiling fans present

When you carry out a spillage test on an open-flued warm air heater
(and circulator where fitted) if there is a re-circulatory ceiling fan fitted in the same room, carry out spillage tests:

- with the fan both on and off;
- at all speeds; and
- where appropriate, with the fan operating in both directions

Tests indicate that these fans can disturb air movement in the room - to the extent that they can cause spillage to occur where none was present with the fan in the 'off' position.

- if you have any doubt regarding the safe use of the appliance, follow the current Gas Industry Unsafe Situations Procedure (see also the current Essential Gas Safety – Domestic – Parts 8 and 10 for further guidance)

How to carry out servicing

A warm air heater must be correctly serviced to:

- maintain the performance of the appliance; and
- to provide the user with optimum comfort conditions

Complaints of lack of performance (which may make the user want to replace the heater) sometimes simply stem from a poorly serviced circulation fan unit impeller and motor assembly that is restricted with dust/lint etc.

- follow the manufacturer's instructions supplied with the appliance, where available. They are specific to the individual appliance and recommend time between services and instructions on how to access various components within that appliance
- where a circulator is an integral part of the appliance, carry out servicing in accordance with the manufacturer's instructions – or, if they are not available, see the guidance in the current Gas Installer Manual Series – Domestic – 'Water Heaters'

General procedure (for all appliances when applicable)

Always follow the manufacturer's servicing procedure, in the absence of instructions, follow this general procedure for appliances covered in this Part.

Preliminary examination

1. Check with the customer to ascertain any problems with the appliance and/or heating system.
2. Check the location of the appliance is suitable (see **Part 2 – Open-flued boilers** and/or **Part 3 – Room-sealed boilers; Restricted locations – for safety reasons**, as appropriate).
3. Check for any damage to the appliance and surroundings and advise the customer, as appropriate, before you start any work.
4. Check the operation of the appliance, controls, including thermostats, ignition system, FSDs.
5. Check the appliance burner flame picture(s).
6. Where applicable, check the electrical installation complies with the Requirements of BS 7671.
7. Check clearances from combustible materials e.g. compartments etc.
8. Check gas installation pipework for exposure to corrosion, sleeving and contact with electrical cables.

Full service

1. Isolate the appliance from the gas supply and where applicable, the electricity supply (see **Part 8 – General installation details – Wet central heating – Electrical isolation – essential for safety**).
2. Because of the possibility of stray electrical currents, consider attaching a temporary continuity bond to gas supply and appliance (see the current Essential Gas Safety – Domestic – Part 5 for further guidance).
3. Remove the main burner – wherever possible dismantle the burner and clean any internal filter or lint arrester gauze as follows:
   a) Remove all surface dust with a paint brush or similar.
   b) Using a combination of brushes, remove dust and lint from within the primary air ports, venturi and burner(s).
   c) Check the burner(s) for cracks and metal fatigue.
4. Clean main burner injector(s).
5. Remove the pilot assembly – clean the burner and injector.
6. Check the pilot supply tube is clean and unobstructed.
7. Where necessary, access the heat exchanger and thoroughly clean it with a suitable flue brush or tool.
8. Inspect the condition of the heat exchanger and flue ways for signs of metal fatigue (cracking or distortion) – see 'Attention' and **Testing heat exchangers – check for POC** in this Part.

9. Re-assemble the burner(s) and pilot assembly.
10. Check the condition of the ignition lead and the alignment of the electrode.
11. Turn on the gas and test all disturbed joints for gas tightness using non-corrosive LDF.
12. Carefully remove the fan assembly (which generally includes the fan motor) and thoroughly clean the fan blades, taking care not to disturb any balance weights (where fitted).
13. Reassemble and refit the fan assembly and where applicable check and refit the fan belt.
14. Restore the electrical supply.
15. If necessary, adjust the pilot flame, to envelop the thermocouple tip. If the pilot flame is extinguished, do not try to re-light the appliance for at least 3 minutes.
16. Test the FSD (or atmosphere sensing device (ASD) where fitted) for correct operation (see 'Attention' and also the current Essential Gas Safety – Domestic – Part 12 for further guidance).
17. Re-light and check the appliance gas pressure, gas rate or both (where necessary) in accordance with the appliance data plate and adjust as necessary.
18. Check the main burner and pilot for satisfactory flame picture and ensure that they are not disturbed when the air circulation fan is running (see 'Attention' and see also **Testing heat exchangers – check for POC** in this Part).
19. Check the high limit (or overheat) thermostat switch operates correctly - by running the heater with the main burner alight and the fan disconnected. The heater should shut down within 3 to 5 minutes (carry this out following the manufacturer's instructions).
20. Check the heater air filter is clean and unobstructed and renew as necessary (see **Part 9 – Warm air heating – A range of air filters for warm air heaters**).
21. Check the ventilation requirements are correct (see **Ventilation – when and why needed** in this Part and also refer to BS 5440-2 for further guidance)
22. Where applicable, carry out a flue flow test and check the chimney termination is correct (see the current Essential Gas Safety – Domestic Parts 13 and 14 for further guidance).
23. Where applicable, carry out a spillage test (see **Commissioning – Spillage testing** in this Part). See also the current Essential Gas Safety – Domestic – Parts 13 and 14 for further guidance (see Note).
24. Where applicable, check that the room-sealed terminal is installed correctly (fit a terminal guard where necessary) and that no undergrowth will interfere with combustion and adversely affect chimney performance.
25. If room-sealed, check that the appliance case seals are in good condition, and renew any sealing material as necessary. Also make sure the case itself fits securely and that all fixing bolts/screws are located correctly.
26. Inspect all the return air paths and if the appliance is open-flued, check if there is a positive return air connection to the appliance. See 'Attention' and also see **Part 9 – Warm air heating – Return air (re-circulation)** for further guidance.
27. Advise the user to have the appliance(s) serviced/safety checked at a minimum of 12 monthly intervals, or at intervals specified in the manufacturer's instructions.

28. Advise the user of any appliance/installation defects in writing and where necessary, follow the current Gas Industry Unsafe Situations Procedure (see the current Essential Gas Safety – Domestic – Parts 8 and 10 for further guidance).

Attention: If gas appliances are fitted with an ASD pilot, service these devices in strict accordance with the manufacturer's instructions.

- they are not 'field adjustable'. This means that if a fault develops on, for example, the thermocouple lead, then you may need to replace the entire unit
- some manufacturers recommend replacing the ASD every five years (see the current Essential Gas Safety – Domestic – Part 12 for further guidance)

Important: You must closely examine the heat exchanger. Should a crack develop in this, circulation air may be blown into it, causing flame turbulence - particularly when the fan switches on (see 'Testing heat exchangers – check for POC' in this Part).

Flame turbulence can often occur because the heater is not properly sited or sealed onto the plenum. This allows air to escape into the heater compartment.

A build-up of pressure can then cause flame turbulence similar to that caused by a heat exchanger fault.

- if flames are disturbed, further investigate the heat exchanger condition and/or position and sealing of the heater onto the plenum

Attention: For existing installations, where there is no positive return air connection, class the installation as 'At Risk' (AR) in line with the current Gas Industry Unsafe Situations Procedure (see also the current Essential Gas Safety – Domestic – Parts 8 and 10 for further guidance).

- if there is no provision on the appliance to install a positive return air connection, seek advice from the appliance manufacturer, or other warm air specialist

Note 1: If any room or premises is fitted with a fan (e.g. decorative re-circulatory ceiling fan, extract fan, or a fan incorporated within an appliance (including warm air heaters and tumble dryers)), operation of the fan(s) must not adversely affect the performance of the chimney when you test the appliance in accordance with the manufacturer's instructions (see the current Essential Gas Safety – Domestic – Part 14 for further guidance).

Note 2: There are no specific instructions for servicing a multifunctional gas control valve – but be aware that when you depress the pilot control knob to establish the pilot flame, the control knob should be free and easy to operate.

If this is not the case:

- remove the plastic knob and apply a small amount of the control manufacturer's lubricating oil to the spindle

Failure to correct this fault could lead to a serious gas escape on the control

Cleaning heat exchangers – essential for safety

Some warm air heater heat exchangers have flue ways that are difficult to reach and clean.

- when you service such appliances, particularly if the heat exchanger is full of soot, make sure that all flue ways are clean
- if you cannot remove accumulated soot, you may need to remove the heat exchanger from the appliance to adequately clean it
- if you fail to effectively clean these flue ways, it will restrict or prevent the POC from passing through the heat exchanger. This will almost certainly result in the heater becoming blocked with soot once again
- during this period the user may be in danger from CO poisoning, especially from open-flued appliances

Testing heat exchangers – check for POC

Because of the possibility of cracking/leakage, with the prospect of POC entering the property that could contain CO, check the integrity of the warm air heater heat exchanger when you service the appliance, in the absence of manufacturer's instructions, use the following test methods:

Method 1

1. Isolate the gas and electrical supplies to the appliance.
2. Where possible, remove the front panel(s).
3. Remove the controls and burner assemblies.
4. Disconnect and remove the air circulation fan assembly.
5. Using a powerful torch through the fan aperture, examine the heat exchanger externally (looking for cracks and holes) - paying particular attention to welded joints. Also check the chimney is fitted correctly to the top of the heat exchanger.
6. Again with the aid of a powerful torch, examine the heat exchanger internally (looking for cracks and holes) paying particular attention to welded joints.
7. With the torch still positioned inside the heat exchanger (being used as a light source), again examine the heat exchanger externally through the fan aperture, in the same places as before, but this time looking for signs of light.
8. Following re-assembly, turn on and re-light the appliance, check and where necessary, adjust the burner pressure/gas rate.
9. Visually inspect the flame picture. If the flame picture is disturbed when the warm air circulation fan is running, check for any air leaks between the heater and the plenum, paying particular attention to appliances with a rear down draught diverter. Leakage may also stem from a poorly fitted chimney connection. Rectify any defects/air leaks before you continue with the procedure.
10. Allow the appliance to achieve normal operating temperature (usually 10-15 minutes) and re-check the stability of the burner flame picture, making sure the warm air circulation fan doesn't affect flame stability.
11. If you find no defects and the appliance is operating correctly, complete commissioning/servicing and put the appliance back into operation.
12. If you find defects, see 'Attention' to Method 2.

Method 2

The smoke pellet is another test medium commonly used to check the integrity of the heat exchanger (see Note).

1. Light the appliance and allow it to heat up to normal operating temperature (generally 10-15 minutes).
2. Turn off the appliance (gas and electricity).
3. Close all system registers except the one closest to the appliance.
4. Place a lighted smoke pellet into the combustion chamber (towards the rear) and allow it to burn out.
5. Switch on the appliance circulation fan and check for traces of smoke at the remaining open register.
6. Where you encounter smoke, visually check the integrity of the heat exchanger and chimney again (see 'Attention').
7. If you find no defects, put the appliance back into operation.
8. Turn on and re-light the appliance, check and where necessary, adjust the burner pressure/gas rate.
9. Visually inspect the flame picture. If the flame picture is disturbed when the warm air circulation fan is running, check for any air leaks between the heater and the plenum, paying particular attention to appliances with a rear down draught diverter. Leakage may also stem from a poorly fitted chimney connection. Rectify any defects/air leaks before you continue with the procedure.
10. Allow the appliance to achieve normal operating temperature (usually 10-15 minutes) and re-check the stability of the burner flame picture, making sure that the warm air circulation fan does not affect flame stability.
11. If you find no defects and the appliance is operating correctly, complete commissioning/servicing and put the appliance back into operation.
12. If you find defects, see 'Attention'.

Attention: Whichever method you choose, if you find the heat exchanger of the appliance is perforated, regard it as 'Immediately Dangerous' (ID) – and follow the current Gas Industry Unsafe Situations Procedure (see also the current Essential Gas Safety – Domestic Parts 8 and 10 for further guidance).

Note: Method 2 has its limitations –

- you must ensure the size of the smoke pellet you use is not too great in relation to the amount of smoke it produces

In the case of an open-flued appliance, the smoke produced may be greater in volume than the heat exchanger and chimney of the appliance can handle. This can cause smoke to 'spill' from the burner or down draught diverter positions, giving a false result

Once the smoke has 'spilt' into the area surrounding the appliance, it's difficult for you to detect whether or not there were any leaks within the heat exchanger in the first place

- with Method 2, it is also recommended you visually inspect the heat exchanger to confirm its integrity, especially in the area of the combustion chamber

Maintenance

Where you carry out any maintenance work on a gas appliance e.g. clearing a blocked pilot jet etc. the Gas Safety (Installation and Use) Regulations require you to examine:

1. The effectiveness of any flue.
2. The supply of combustion air.
3. Its operating pressure or heat input or, where necessary, both.
4. Its operation so as to ensure its safe functioning.

You must then take all reasonable practicable steps:

- to notify any defect to the responsible person and where different, the owner of the premises in which the appliance is situated; or
- where neither is reasonably practicable, in the case of an appliance supplied with Liquefied Petroleum Gas (LPG) the supplier of the gas to the appliance or, in any case, the transporter

Fault finding – always be methodical

The operation of appliance, burners, control taps, ignition systems, thermostatic controls and FSDs are covered in detail in the appropriate parts of the current Essential Gas Safety – Domestic – Parts 1-17.

However, the following list helps you be methodical when fault finding.

General fault finding guide to help you

1. Check with the customer to ascertain the particular problems they have with the appliance. This helps to pin point any defects.
2. Check the location and general installation requirements for the appliance are in accordance with the manufacturer's installation instructions.
3. Where possible, always refer to the appliance manufacturer's installation/maintenance instructions. They often contain fault finding information including flow charts to guide you to a satisfactory conclusion. They may also contain specific information about testing and replacing particular parts.

If an electrical fault occurs after the warm air heater is installed, a competent person must carry out a preliminary electrical system check once you have completed any fault finding task where you have had to break and remake electrical connections, you must carry out checks on these connections for:

- continuity
- polarity
- resistance to earth

Take care when replacing and handling electrical components, faults in some assemblies can only be rectified in the factory - and if you attempt to do so, it may render any guarantee or factory replacement void.

Note: Many faults occur on electrical components. You must be competent to carry out these checks.

- the fault finding charts (Tables 11.1 and 11.2) only cover basic checks needed. You may find a more comprehensive fault finding chart in the appliance manufacturer's instructions

11.1 Fault finding chart – basic model

Symptom	Possible cause	Action
Pilot will not light	No gas to appliance.	Check main gas supply.
	Gas supply pipe not purged.	Purge. See the current Essential Gas Safety – Domestic – Part 15 for guidance
	Pilot injector restricted.	Carefully clear injector or replace.
	Faulty electrode or cable.	Replace.
	Faulty igniter.	Replace.
Pilot light goes out on releasing 'Start' button during initial light up or after normal operation.	Connection between thermocouple and gas valve loose.	Check connection and secure.
	Faulty power unit on gas valve.	Replace.
	Faulty thermocouple.	Replace.
	Pilot flame does not envelop thermocouple.	Adjust as necessary.
Pilot lit but main burner not igniting.	Mains electric supply not connected to heater.	Check main electrical supply.
	Fuse failed.	Replace. If fault occurs again, check room thermostat leads for short to earth with electrical test meter.
	Controls not calling for heat.	Check time control and/or room thermostat is calling for heat.
	Loose connection on room thermostat, limit control, gas valve, time control or transformer.	Check connections.
	Transformer open circuit.	Check with electrical test meter, replace if necessary.
	Gas control solenoid operator faulty.	Replace.
	Gas control regulator faulty.	Replace.
	Limit control faulty.	Check with electrical test meter, replace if necessary.
	Faulty room thermostat or external wiring.	Check with electrical test meter, replace if necessary.
	Heater controls set on 'summer' setting.	Reset controls.
Main burner lights but fan fails to run after preheat period.	Loose electrical connection on fan control.	Check connections.
	Fan switch settings incorrect.	Check settings.
	Fan switch faulty.	Replace.
	Faulty fan assembly.	Replace.
	Burner pressure setting incorrect.	Check setting and adjust.
	Fan belt (if fitted) faulty.	Adjust or replace.

11.1 Fault finding chart – basic model (continued)

Symptom	Possible cause	Action
Main burner operating intermittently with fan running.	Gas rate or burner setting pressure too high.	Check settings and adjust.
	Temperature rise excessive.	Adjust fan speed.
	Air filter or return air path restricted.	Check filter is clean and path is clear.
	Most outlets (registers/diffusers) closed.	Open additional outlets (registers/diffusers).
Fan operating intermittently with main burner lit.	Gas rate or burner setting pressure too low.	Check settings and adjust.
	Fan switch setting incorrect.	Check settings.
Fan runs for excessive period or operates intermittently after main burner shuts down.	Fan switch setting incorrect.	Check settings.
Noisy operation.	Gas pressure too high.	Check burner setting pressure.
	Noisy fan motor/unit.	Clean fan blades. If still noisy, check the condition of the fan/motor mountings/bearings/alignment of the belt/pulley. Where necessary, replace fan motor/unit.
Main burner remains ON with controls set to OFF.	Multifunctional gas valve fails to close down.	Disconnect or turn off electrical supply to valve. If burner does not close down, replace valve.

11.2 Fault finding chart – Even Temperature (ET) models only

Symptom	Possible cause	Action
Main burner not lighting but pilot is alight, voltage detected across gas control valve.	Gas pressure regulator set too low.	Adjust regulator.
	Multifunctional control solenoid operator faulty.	Replace operator.
Main burner not lighting but pilot alight, voltage not detected across gas control valve.	Fault in mains electrical supply.	Check electrical supply.
	Internal fuse blown.	Replace fuse.
	Replacement fuse blows due to fault in gas control solenoid operator.	Disconnect wires to gas control and check with electrical test meter, check connections and replace solenoid operator if necessary.
	Replacement fuse blows due to fault in electronic panel.	Replace electronic panel.
	Fault in external wiring to thermistastat either: a) Break in circuit. b) Reversed polarity.	a) Check for continuity by bridging at thermistastat plug. Main burner should light. b) Check for correct polarity at thermistastat terminal block.
	Faulty thermistastat.	Replace thermistastat.
	Faulty limit control.	Check with electrical test meter, check connections and replace limit control if necessary.
	Faulty electronic panel.	Replace electronic panel.
Main burner lights but fan fails to run even when override switch is set to continuous.	Poor electrical connection on fan circuit.	Check connections, especially plug and socket.
	Faulty fan assembly.	Replace fan assembly.
	Faulty electronic panel and/or fan speed regulator.	Replace electronic panel and/or fan speed regulator.
Main burner lights but fan fails to run when override switch is set to AUTO from CONTINUOUS.	Faulty electronic panel or fan speed regulator.	Replace electronic panel and/or fan sped regulator.
	Faulty air flow sensor.	Bridge across airflow sensor. If fan runs replace sensor.
Main burner operates for short periods only on initial light up.	Fault in external wiring to thermistastat, either: a) Break in circuit. b) Reversed polarity.	a) Check continuity by bridging wires at thermistastat plug. b) Check for correct polarity.
Main burner remains ON with controls set to OFF.	Multifunctional gas valve fails to close down.	Disconnect or turn off electrical supply to valve. If burner does not close down, replace valve.

Odd coloured gas flames

Some warm air heating appliances covered by this Part have burner flames that are visible and exposed to the air in the dwelling. This does not generally present a problem.

However, if a householder with a respiratory condition uses a nebuliser to relieve the symptoms, the gases given off by this may affect the flame characteristics and hue.

Depending on the concentration of these gases in the room, this may cause the burner flames to change colour - varying from a pale pink to a bright orange. The flames can also appear to be much larger than normal, as salts in the gases expose the full outer mantle of the flame, which is normally not visible to the naked eye.

The flames will return to their normal characteristic size and colour once the room(s) affected by the gas from the nebuliser are purged with fresh air.

Chimneys

Open-flues - requirements

- only use chimneys and chimney fittings complying with BS EN 1856-1
- install the chimney installation according to the manufacturer's installation instructions

Detailed guidance on the installation of chimneys and termination positions will be found in the current Essential Gas Safety – Domestic Part 13.

- before you install a warm air heating appliance to an existing chimney system, carry out a flue flow test to verify the chimney is operating correctly
- when you have completed the installation, carry out a spillage test in accordance with the manufacturer's instructions to ensure that the POC are being safely removed (see the current Essential Gas Safety – Domestic Part 14 for further guidance and also **Commissioning: your responsibility** in this Part)
- before you install any gas appliance, thoroughly sweep any chimney previously used for an appliance burning a fuel other than gas

Note: Follow manufacturers' instructions when you install combined warm air heaters/circulators to open-flues.

Poured/pumped concrete chimney liners – checks needed

- these are acceptable alternatives to flexible metallic liners, but must only be dealt with by a competent contractor
- install poured/pumped concrete chimney linings using a method certificated by an accredited test house

- always check the lining is mechanically sound before you install any appliance
- before use with a gas appliance, sweep and carefully examine chimneys with this type of lining that have been used with another fuel

Flue terminals – when to fit

- fit an approved terminal if a brick/masonry chimney has been lined with a rigid metal chimney system or flexible metallic liner - or where an appliance has a direct flue connection (see the current Essential Gas Safety – Domestic – Part 13 for further guidance)
- fit a chimney of 170mm diameter or less across the axis of its outlet with a terminal. The size of the terminal should not be less than the nominal size of the appliance flue connection (see the current Essential Gas Safety – Domestic – Part 13 for further guidance)

Flue testing

- when you plan installing an open-flued warm air heating appliance to an existing chimney system, verify its correct operation (see also the current Essential Gas Safety – Domestic – Part 14 for further guidance)
- when installation and commissioning is complete, test the appliance and chimney in accordance with the manufacturer's instructions - to ensure that POC are not spilling into the room (see the current Essential Gas Safety – Domestic – Part 14 for further guidance)

  However, some warm air appliance manufacturers have spillage testing instructions specific to a particular model. If so, this information will be in their installation instructions (see also **Commissioning – Spillage testing** in this Part).

Natural draught room-sealed warm air heaters

- choosing the terminal position on the outside wall is probably the most critical part of the installation

For gas to burn correctly (complete combustion) POC should readily disperse and pass freely away from the concentric flue terminal (combined air inlet and flue outlet duct) into the atmosphere and not mix with clean fresh air passing through the air inlet duct to the burner.

- when you install a heater to harmonise with kitchen units, you must also consider the flue outlet position. It must not be restricted by an adjacent projection outside - e.g. buttresses, gate posts, soil pipes, internal or external corners of buildings etc known as 're-entrant positions'
- you must ensure that the POC are prevented from being blown away by the wind, re-circulating around the terminal and 're-entering' the appliance via the fresh air inlet duct

The POC vitiate this fresh air and reduce the oxygen content.

Consequently, this has a profound effect on combustion quality and is likely (depending on the degree of vitiation) to cause the flames, including the pilot flame, to become ragged and lift off the burner. In the case of the pilot burner, this could eventually lead to cooling of the thermocouple, which in turn could cause the appliance to fail to safety.

The condition is particularly aggravated on windy days (see the current Essential Gas Safety – Domestic – Part 13 for further guidance).

- in the examples given, regard the condition as unsafe
- when you encounter it, follow the current Gas Industry Unsafe Situations Procedure (see also the current Essential Gas Safety – Domestic – Parts 8 and 10)

Similar conditions occur if the concentric flue and air inlet duct is cut too short for the wall thickness. The air inlet grilles are then likely to be obstructed or blocked by cement mortar, which restricts air entrainment and adversely affects combustion.

Other 're-entrant' positions are openings into buildings such as doors, windows and ventilators.

- some appliance manufacturers now stipulate particular dimensions where you must site flues away from openings into buildings. Always comply with these
- if there are no particular instructions, seek guidance from the appliance manufacturer
- it is also important that you position the terminal so that POC can safely disperse at all times

For example, when the termination is into a car-port or other similar structure, there should be at least two open unobstructed sides to that structure.

- pay attention to the material used on the roof and provide allowance for adequate clearances/protection of the roof
- don't site terminals into a passageway, pathway or over adjoining property where they can be a nuisance or cause injury (see the current Essential Gas Safety – Domestic – Part 13 for further guidance)
- some natural draught room-sealed warm air heaters are also suitable to install onto 'Se-duct' or 'U-duct' chimney systems

Fanned draught room-sealed warm air heaters

Installation requirements are generally the same as those for natural draught room-sealed heaters, though the siting of a terminal for a fanned draught warm air heater is not so critical.

This is because the fan assists with dispersal of the POC - thus eliminating most of the problems associated with the siting of natural draught room-sealed terminals.

- whilst the siting requirements are more relaxed, take the same precautions as for natural draught room-sealed flue terminations e.g. when the termination is close to openings into buildings or is into a car-port or other similar structure
- ensure that the POC and any pluming are not blown onto an adjacent property, causing a nuisance (see the current Essential Gas Safety – Domestic – Part 13 for further guidance)

These warm air heaters are generally designed to incorporate several flueing options with side, rear or vertical flue outlet positions available.

Gas supply: your responsibility

- the Gas Safety (Installation and Use) Regulations require that when you install gas installation pipework and fittings, you must make sure you install them safely
- you must give due regard to the position of other pipes, pipe supports, drains, sewers, cable, conduits and electrical apparatus and to any parts of the structure of any premises in which it is installed, which might affect its safe use
- after connecting the appliance, you must test the installation for gas tightness and purge all installation pipes through which gas can flow of air (see the current Essential Gas Safety – Domestic – Parts 6 and 15 for further guidance)
- you will find additional information on LPG tightness testing in the current Gas Installer Manual Series – Domestic – 'LPG – Including Permanent Dwellings, Leisure Accommodation Vehicles, Residential Park Homes and Boats'

Note: Guidance on the installation of pipework/fittings and pipe sizing is in the current Essential Gas Safety – Domestic – Part 5.

Ventilation – when and why needed

All open-flued gas appliances need air for combustion and to assist the safe operation of the chimney system.

Gas appliances up to 7kW heat input (gross or net) generally don't require additional ventilation to be provided - but rely for their correct operation and that of the chimney on adventitious ventilation to the room.

You will find guidance on ventilation, adventitious ventilation, compartment ventilation etc. and the installation of ventilation grilles and sizes in the current BS 5440-2 and the current Essential Gas Safety – Domestic Part 4.

Warm air heaters supplied with combustion air from a ventilated roof space

In today's modern, air tight constructed dwellings (which often include draught and sound proofing material) an alternative method of providing air for combustion is by drawing air from the roof space into the warm air ducting system.

This eliminates the need for openings in walls, doors or windows - and reduces the risk of draughts.

Depending on the appliance manufacturer's particular requirements, the construction of the dwelling and the position of the warm air heater, it may help you to use the following fan assisted method of introducing air for combustion:

1. Connect a duct (fitted with a bird guard) from a ventilated roof space, or from a waterproof grille on an outside wall, to the return air duct or return air plenum on the warm air heater (see Figure 11.2 overleaf).

Figure 11.1 Fanned air supply

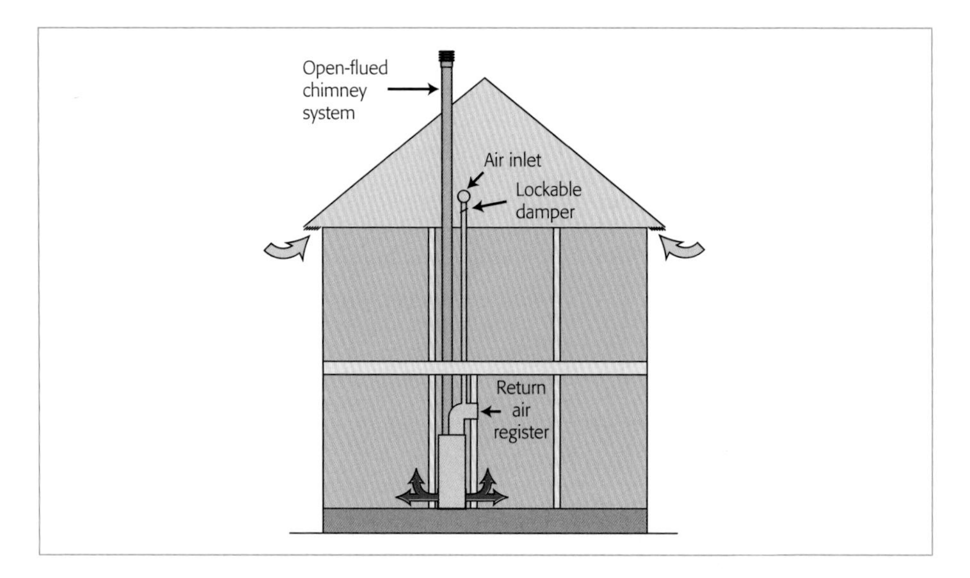

2. Install a lockable damper in the fresh air duct - and adjust to provide the necessary volume of air. A minimum airflow rate of $2.2m^3$ per hour should be drawn into the air chamber for every 1kW of the appliance maximum input rating (net). You should carry out this adjustment. When commissioning is complete, lock the damper in position so that no unqualified person can make an adjustment.

3. When the heater operates, its fan will draw in external air, mix it with the return air and circulate it throughout the warm air distribution ducts. Warm air should be circulated into the room/space where the heater is installed, using a non-closing register fitted in that area.

Note: If you use this method of air supply, refer to the appliance manufacturer's installation instructions for further guidance.

Controls particular to warm air heaters

Room thermostat

For basic fan assisted models a 24V room thermostat is essential to ensure close control and comfort conditions. A heat anticipator is located within the thermostat and is graded in amps (A).

- check the heat anticipator and, where necessary, adjust to correspond with the amp rating of the multifunctional gas control valve (normally set to 0.2 A)
- the room thermostat brings on the main burner when it is calling for heat, provided any time control is set to an 'on' period. The thermostat controls the warm air system output from the air temperature it senses, so it is most important you select its optimum location

Two main choices for you to locate it are:

1. In the living room, where the temperature has the most significant effect on comfort, or
2. Close to the main return air grille. This position is preferable for systems that are likely to be used continuously to heat all rooms.

Mount the thermostat:

1. At a height of 1200mm to 1500mm above floor level.
2. On an inside wall.
3. Away from direct sunlight and other local heat sources e.g. space heating appliances, wall light fittings or television sets.
4. Away from warm air ducts, diffusers, registers or the heater itself.
5. Away from outside walls, doors, or windows.

On modern warm air heating systems, the room thermostat and the programmer control the main gas valve. The fan is controlled by a series of other controls.

Overheat control required

If air is not circulating through the heat exchanger of a warm air heater, this can quickly become overheated.

- so an electronic cut-off device is needed to shut off the gas supply to the main burner, to prevent overheating occurring. The maximum temperature is set at 110°C
- to prevent overheating, you can use an overheat or a limit thermostat (see **Limit thermostat** in this Part) or in some cases; both
- the overheat is usually provided as an additional control for down-flow heaters because, in these models, the filter assembly at the top of the heater can quickly rise to a high temperature if the fan assembly is slow in switching on

Limit thermostat

This is similar to the overheat control and on many warm air heaters is the only temperature limiting device fitted.

The limiting temperature is set at a maximum 95°C. The bimetal-switching device is wired in series with the gas control valve. The limit thermostat has a differential setting of about 15°C. So the heater will be turned on again when the temperature falls to about 80°C. Because the limit thermostat is automatically reset, the heater will cycle on and off every few minutes when an overheat condition occurs.

Fan switch

If the fan and the gas control valve were both switched on at the same time (as many old warm air heaters operated in the past), the fan blew cold air into the rooms before the heat exchanger had time to heat up.

So modern heaters are fitted with a thermally operated switch to delay the fan operating until the heat exchanger has reached a pre-determined temperature. It's normally operated by a bimetal-switching device located in the outlet of the heat exchanger and is wired in series with the fan.

Usual temperature settings are:

- fan 'off' at about 38°C
- fan 'on' at about 58°C; or
- 'differential' of about 20°C

Operating the fan switch will switch on the fan when the air in the heat exchanger has reached about 58°C. When the gas solenoid has closed, the fan will continue running until the temperature of the air leaving the heat exchanger falls to about 38°C.

This fully utilises the residual heat from the appliance.

Fan delay unit: another type of fan switch

This is a micro-switch operated by a bimetal strip which is heated by a resistor. The resistor is in series with the room thermostat, in the low voltage circuit and the micro switch is in the 230V a.c. supply to the fan.

When the programmer is switched on and the room thermostat is calling for heat the gas control valve becomes energised and the fan delay unit resistor begins to heat up. After about 90 seconds, the bimetal strip operates the micro-switch and switches on the fan.

When the room reaches the required temperature, the room thermostat switches off and the supply to the resistor is cut off. The fan then continues to run for a short period while the resistor cools down.

Summer/winter switch

This is a simple, manually controlled switch, fitted in the main voltage supply to the fan. If the user wishes to run the fan in the summer to circulate cool air through the rooms (when the heating unit is not operating), the switch is changed to the 'summer' position.

Even temperature (ET) controls - use thermistastat

Fan assisted warm air heaters with basic controls are generally operated by a 24V thermostat.

When heat is needed, it operates an electrical panel to bring the main burner on at a pre-selected gas rate. The fan switch then brings on the air circulation fan at a single pre-selected speed. Warm air is circulated and delivered through the ductwork system until the room thermostat is satisfied. The heater then simply shuts down.

Draughts and noise can be a problem on short ducted systems, and there can be a significant temperature difference across rooms and spaces. So some manufacturers have designed modulating gas burners and/or air circulation fans to give more even temperature control to overcome this problem.

They are operated by an electronic controller called a 'thermistastat' which is fitted instead of a room thermostat.

How the thermistastat works

It is a heat-sensitive resistor which sends a continuous signal to the heater and you locate it as you would for a room thermostat.

The thermistastat senses the heat requirements and continuously sends signals of varying strengths to the electronic controls in the heater to control the rate of warm air delivery. The warm air heater controls automatically adjust the operation of both the gas burner and the air circulation fan. The main burner cycles at approximately 2 minute intervals and the fan speed is then matched to the heater output.

Definitions – 13

Definitions

'R' type adaptor: see **Ridge terminal adaptor ('R' type adaptor) in this Part**.

1st family gas: at present normally only LPG-air.

2nd family gas: Natural gases.

3rd family gas: Liquefied Petroleum Gases (LPG).

Access plate: a removable plate giving access into a combustion chamber and/or heat exchanger.

Adventitious ventilation: ventilation obtained through gaps around doors, floors and windows, for example.

Aerated burner: a burner in which some or all of the required air has been mixed with the gas before it leaves the burner port(s).

Air heater: appliance for heating air to be used for space heating.

Air vent free area: total area of the unobstructed openings of an air vent.

Air vent: non-adjustable grille or duct, which allows the passage of air at all times.

Anemometer: instrument for recording wind speed.

Annular space: space between a flexible metallic flue liner and brick/masonry chimney.

Appliance (gas): appliance designed for heating, lighting, cooking or other purposes.

Appliance compartment: an enclosure (not being a habitable space) specifically designed or adapted to house one or more gas appliances (see also Balanced compartment in this Part).

Appliance ventilation duct: provided to convey combustion or cooling air for an appliance or component.

Atmosphere sensing device: (also known as an oxygen depletion system) shuts off the gas supply to an appliance burner before there is a build up of a dangerous quantity of products of combustion in the room concerned.

Background central heating: simultaneous space heating to temperatures below those specified for full central heating.

Balanced compartment: method of installing an open-flued (Type B) appliance as room-sealed, so that flueing and ventilation provide a balanced flued effect.

Balanced flued appliance: a room-sealed appliance (Type C) which draws its combustion air from a point adjacent to that at which the products of combustion are discharged, the inlet and outlet being so disposed that wind effects are substantially balanced.

Basement (LPG appliances): a room, which is completely or partly below ground level on all or some sides.

Bedsitting room: any room or space used for living and sleeping purposes.

Boiler: appliance designed to heat water for space heating and/or water supply.

Boot: a transition from the round or rectangular duct to the diffuser/register.

Bottled gas: normally Butane or Propane stored as a liquid under pressure in refillable portable containers.

Branched flue system: a shared open-flued system serving appliances situated on two or more floors.

Bulk storage vessel: permanently installed vessel of approved design for the storage of LPG under pressure, which is filled on site.

Calorific Value (or CV): the Calorific Value is the quantity of heat (energy) produced when a unit volume of the fuel, measured under standard conditions of temperature and pressure, is burned completely in excess air.

A distinction is made between:

a) Gross Calorific Value – (also known as the Higher Calorific Value or HCV) – in the determination of which the water vapour produced by the combustion of the fuel is deemed to have been condensed into a liquid phase at the standard temperature and its latent heat released.

b) Net Calorific Value – (also known as the Lower Calorific Value or LCV) – in the determination of which water vapour produced by the combustion of the fuel is assumed to remain in the vapour phase. The net Calorific Value is therefore the gross Calorific Value minus the latent heat of the water vapour contained in the combustion products.

Capacity (of a gas meter): maximum rate that gas will flow through the meter, measured in m^3/hr or ft^3/hr.

Central heating system: a fixed system for warming a building from a single source of heat, with manual or automatic control of the operation of the whole system and of the temperatures in the heated space.

Chimney liner: pipe inside a brick/masonry chimney to form a flue. May be rigid or flexible.

Chimney pot: prefabricated unit fitted at the outlet of a chimney.

Chimney system: a complete assembly of chimney components from one or more appliances to a single terminal, including primary flue(s) and draught diverter(s), if any.

Circulation pipe: pipe forming part of the primary circuit of a hot water system.

Circulator: boiler with a rated heat input not exceeding 8kW (gross), designed primarily for the supply of domestic hot water in conjunction with a separate hot water storage vessel.

Cistern: a fixed container for holding water at atmospheric pressure.

Cold feed pipe: the pipe from the feed and expansion cistern to the water heating system.

Collection area: where all heated rooms/spaces are connected to one area by means of a number of air relief openings.

Combination hot water storage unit:

1) a hot water supply apparatus comprising of a hot water storage vessel with a cold water feed cistern immediately above it, the two being fabricated together as a compact unit.

2) a hot water supply apparatus comprising a hot water storage vessel with a cold water feed cistern beside it or inside it.

Commissioning: initial start-up of an installation to check and adjust for safe and reliable operation.

Competence: competence in safe gas installation requires gas operatives to have enough knowledge, practical skill and experience to carry out the job in hand safely, with due regard to good working practice. Knowledge must be kept up-to-date with awareness of changes in law, technology and safe working practice.

Condensate drain: a device in a flue where condensate can be removed.

Condensing appliance: designed to use latent heat from water vapour in the combustion products by condensing the water vapour within the appliance.

Damper: a device used to vary the volume of air passing through a confined cross-section by varying the effective cross sectional area.

Data plate: a durable, permanently fixed plate bearing specified information relative to the appliance.

Designed heat loss: the heat loss from a building estimated from considerations of the structure and the intended working temperature(s) and ventilation rates.

Dew point: the temperature of a mixture of combustion products and water vapour at which further cooling results in condensation of the water vapour.

Diffuser: a fitment equipped with a damper or moveable louvers that permit adjustment or closure of an opening from which air discharges. Generally fitted in a floor or ceiling.

Direct hot water storage vessel: storage vessel with no internal heat exchanger, heated directly by an appliance containing the same water.

Double-feed indirect hot water storage vessel: any indirect hot water storage vessel that requires a separate feed cistern to both the primary and the secondary circuit.

Draught brake: an opening into any part of an open-flue chimney system, including that part integral with the appliance.

Draught diverter: prevents interference to the combustion of an open-flued appliance; must be fitted to the manufacturer's instructions and in the same room, space or compartment as the appliance, with at least 600mm of vertical chimney above it.

Draw-off point: hot water taps.

Drop-out time: time taken for the flame supervision system to respond to a loss of flame.

Ducted warm air heater: flued appliance, which uses ducts to distribute the heated air.

Emergency control valve (ECV): valve for shutting off the supply of gas in an emergency; not a service isolation valve (see also Additional emergency control valve (AECV) in this Part).

Enforcing authority: an authority with a responsibility for enforcing the Health and Safety at Work Etc. Act 1974 and other relevant statutory provisions; normally Health and Safety Executive (HSE) or the local authority for the area as determined by the Health and Safety (Enforcing Authority) Regulations 1977.

Equipotential bonding: electrical conductor between a point close to the outlet of a gas meter and the earth terminal (such bonding does NOT involve connecting electrical power to gas pipework).

Fanned draught flue system: flue system in which the draught to remove products of combustion is produced by a fan.

Fanned draught room-sealed appliance: an appliance that, when in operation, has the combustion system including the air inlet and the products of combustion outlet, isolated from the room or space in which the appliance is installed. The draught to operate the flue is created by an integral fan.

Feed and expansion tank: cistern which supplies cold water to the primary circuit of a heating system which allows the expansion of the system water when hot.

Fire compartment: room or space constructed to prevent the spread of fire.

Fire stop: a barrier or seal of non-combustible material that is designed to prevent or retard the passage of smoke or flames.

Flame failure: the loss of a flame from the normally detected position.

Flame retention: prevention of flame-lift off.

Flame supervision device: control, which detects the presence of a flame and in the absence of that flame, prevents the uncontrolled release of gas to the burner.

Float (ball) – operated valve: a valve for controlling the flow of water into a cistern.

Flow pipe: a pipe in a primary hot water circuit in which water moves away from a circulator.

Flue break: an opening in the secondary flue in the same room as and in addition to, the opening at the draught diverter.

Flue outlet: the part of the appliance that allows the exit of products of combustion from the appliance.

Flue safety device: a device designed to detect adverse flue conditions (down draught) at the appliance draught diverter; also known as a TTB.

Flue terminal guard: fitted to prevent human contact (especially that of children) with a terminal and to prevent interference with the terminal or damage to it.

Flue terminal: device fitted at the flue outlet to:

a) Assist the escape of products of combustion.

b) Minimise downdraught.

c) Prevent flue blockages.

Flue termination: the outlet of a chimney system where products of combustion discharge into external air.

Flue: passage for conveying the products of combustion from a gas appliance safely to atmosphere.

Free area: total area of the individual unobstructed openings of an air vent.

Gas fittings: "Gas fittings" means gas pipework, valves (other than emergency controls), regulators and meters and fittings, apparatus and appliances, designed for use by consumers of gas for heating, lighting and other purposes, for which gas can be used (other than the purpose of an industrial process carried out on industrial premises), but it does not mean:

a) any part of a service pipe.

b) any part of a distribution main or other pipe upstream of the service pipe.

c) a gas storage vessel.

d) a gas cylinder or cartridge designed to be disposed of when empty.

Gas meter: an instrument for measuring and recording the volume of gas that passes through it without interrupting the flow of gas.

Gas Safety Regulations: legally binding requirements for safe gas work.

Indirect hot water storage vessel: any hot water storage vessel in which the stored hot water is heated by the primary heater through which hot water is circulated from a circulator without mixing of the primary or secondary water taking place.

Individual chimney system: chimney system, which serves a single appliance only.

Insulated chimney – factory-made: complete assembly of all essential factory-made insulated sections, fittings and accessories to convey the products of combustion to the outside air.

Latent heat of condensation: the quantity of heat removed, at constant temperature during the change from the gaseous to liquid state.

Latent heat: the quantity of heat (energy) added to or removed from, a substance at constant temperature during a change of state.

Lint arrester: a fibre pad or metallic mesh designed to trap a mixture of dust, fluff, fibres and droplets of grease which would otherwise collect in the airways of a burner.

Liquefied Petroleum Gas (LPG): normally commercial Propane or Butane gas, stored in a vessel under pressure, which turns into a liquid state.

Lock-out: a safety shut-down condition of a control system such that restart cannot be accomplished without manual intervention.

Main flame: flame on the main burner.

Manometer: instrument(s) for the measurement of gas pressure. Includes 'U' and electronic gauges.

Manufacturer's instructions: documents supplied with the gas appliance/equipment by the manufacturer giving guidance on how to use, service, maintain and install the product.

Mechanical ventilation: air supplied by a fan.

Meter regulator: a device located in close proximity and upstream of a primary meter which is used solely to control the pressure of the gas within the gas installation.

Non-combustible material: that which has passed tests for non-combustibility in accordance with British Standards.

Open vent pipe: a pipe connected to an open water system communicating with the atmosphere.

Open-flued chimney system: system that is open to a room or internal space at each appliance.

Open-flued appliance: an appliance designed to be connected to an open-flued chimney system which draws combustion air from the room or space in which it is installed.

Oxygen depletion system: shuts off the gas supply to an appliance burner before there is a build up of a dangerous quantity of products of combustion in the room concerned, also known as an atmosphere sensing device.

Pluming: visible cloud of products of combustion from an outside flue terminal, which are cooled to below dew point by mixing with external air.

Pre-aerated burner: a burner to which gas and air are supplied already mixed.

Pressure gauge: instrument for indicating or recording pressure.

Pressure regulator: automatically maintains a constant outlet pressure.

Pressure test point: small plug type fitting on a meter, installation pipe or appliance, allowing attachment of a pressure gauge.

Primary circuit: a circuit in which water circulates between a circulator and a hot water storage vessel.

Protected shaft: a shaft which enables persons, air or objects to pass from one compartment to another, enclosed within a fire-resisting construction.

Protected stairway: a stairway including an exit passageway leading to its final exit, enclosed within a fire-resisting construction (other than any part that is an external wall of a building).

Register plate: a fire-resistant plate used to seal the annular space around the base of a flexible metallic flue liner and brick/masonry chimney (see **Annular space** in this Part).

Register: a fitment equipped with a damper or moveable louvers that permit adjustment or closure of an opening from which air discharges. Generally fitted in a wall.

Responsible person: the occupier of the premises or, where there is no occupier or the occupier is away, the owner of the premises or any person with authority to take appropriate action in relation to any gas fitting therein.

Return air duct: a duct through which air returns to a warm air heater.

Return pipe: a pipe in a primary hot water circuit in which water moves back to a circulator.

RIDDOR: The Reporting of Injuries, Diseases and Dangerous Occurrences Regulations 1995.

Ridge terminal adaptor ('R' type adaptor): a fitting for connecting a circular cross-sectional pipe to the rectangular-section connection of a ridge flue terminal.

Ridge terminal: a terminal designed for fitting at the ridge of a building.

Room-sealed: an appliance that, when in operation, has the combustion system, including the air inlet and the products outlet, isolated from the room or space in which the appliance is installed.

Safety shut-off valve: actuated by the safety control so as to admit and stop gas flow automatically.

Secondary flue: the part of the open-flued system connecting a draught diverter or draught break to the terminal.

Se-duct: a duct serving special room-sealed appliances; it is open at both ends and rises vertically in buildings to bring combustion air in and take products of combustion out.

Sensible heat: the quantity of heat added to, or removed from a substance during the finite temperature change.

Single-feed indirect hot water storage vessel: an indirect hot water storage vessel, which has only one cold feed pipe connection to supply both the primary and secondary circuits.

Sleeve: duct, tube or pipe embedded in the building structure allowing the gas installation pipework to pass through a wall or floor; capable of containing the gas.

Temporary continuity bond: a means of providing electrical continuity on a gas supply for safety reasons.

Thermistastat: thermostat containing a thermistor – containing a semiconductor device having a resistance that decreases rapidly with an increase in temperature – used for temperature control.

Thermostat: a thermally actuated control device for maintaining a desired temperature.

Through-room: any room formed by the removal of an intercommunicating wall between two rooms, or any large room formed by two open plan smaller rooms. The opening/archway present between two smaller rooms may have sliding or intercommunicating doors.

Tightness test: the testing of installation pipes and equipment for escapes from the system.

Transfer grille: non-adjustable fitment in a wall, door, or partition, to transfer air between adjacent rooms and/or spaces.

Transporter: a person other than a supplier, who conveys gas through a distribution main.

U-duct: literally, a u-shaped flue system; combustion air is provided by one limb and special room-sealed appliances are connected to the other. The U-duct ends are open and adjacent.

Valve: device to stop or regulate the flow of gas by the closure or partial closure of an orifice by means of a gate, flap or disc.

Vented hot water storage system: a water storage system that is open to the atmosphere via an open vent pipe.

Ventilation opening: includes any means of ventilation, which opens directly to external air, such as the openable parts of a window, a louvre, airbrick, progressively openable ventilator or window trickle ventilator. It also includes any door, which opens directly to external air.

Ventilation: the process of supplying fresh air to and removing used air from, a room or internal space.

Warning pipe: an overflow pipe so fixed that its outlet, whether inside or outside a building, is in a conspicuous position where the discharge of any water from it can be easily seen.

Work: in relation to a gas fitting this includes any of the following activities carried out by any person, whether an employee or not:

a) Installing or reconnecting the fitting.

b) Maintaining, servicing, disconnecting, permanently adjusting, repairing, altering or renewing the fitting or purging it of air or gas.

c) Where the fitting is not readily movable, changing its position; and

d) Removing the fitting.

Note: work in this context does not include the connection or disconnection of a bayonet fitting or other self-sealing connector.

CORGI*direct* Publications – 14

Gas – Domestic

Manual Series

GID1 Essential Gas Safety (Fifth Edition – Second Revised)

GID2 Gas Cookers and Ranges (Third Edition)

GID3 Gas Fires and Space Heaters (Fourth Edition)

GID4 Laundry, Leisure and Refrigerators (out of print)

GID5 Water Heaters (Second Edition)

GID6 Gas Meters (Third Edition)

GID7 Central Heating – Wet and Dry (Fourth Edition)

GID8 Gas Installations in Timber/Light Steel Frame Buildings (Second Edition – Second Revised)

GID9 LPG – Including Permanent Dwellings, Leisure Accommodation Vehicles, Residential Park Homes and Boats (Third Edition – Second Revised)

GID11 Using Portable Electronic Combustion Gas Analysers for Investigating Reports of Fumes (First Edition)

GID12 Using Portable Electronic Combustion Gas analyser – Servicing and Maintenance (First Edition)

FFG2 Fault Finding – wet central heating systems Domestic (First Edition)

Pocket Series

USP1 The Gas Industry Unsafe Situations Procedure (Sixth Edition)

CPA1 Combustion performance testing – Domestic (First Edition)

SRB1 Ventilation Slide Rule (Third Edition – Third Revised)

GRB1 Gas Rating Slide Rule Natural Gas – Domestic (Second Edition – Revised)

GRB2 Gas Rating Slide Rule LPG (Propane) – Domestic (First Edition)

TTP1 Tightness Testing and Purging (Second Edition – Second Revised)

FFG1 Fault Finding Guide (out of print)

TTG1 Terminals and Terminations (Fourth Edition – Revised)

Design Guide

WAH1 Warm Air Heating System Design Guide (out of print)

Forms

All CORGI*direct's* gas forms carry Gas Safe Register® logo under licence from the HSE.

CP1	Gas Safety Record
CP2	Leisure Industry Landlord's Gas Safety Record
CP3FORM	Chimney/Flue/Fireplace and Hearth Commissioning Record
CP4	Gas Safety Inspection
CP6	Service/Maintenance Checklist
CP9	Visual Risk Assessment of Gas Appliances
CP12	Landlord/Home Owner Gas Safety Record
CP14	Warning/Advice Notice
CP26	Fumes Investigation Report
CP32	Gas Testing and Purging – Domestic (NG)
CP43	Risk Assessment for Existing Chimney Systems in Voids Where Inadequate Access for Inspection is Provided

Labels

CP3PLATE	Chimney/Hearth Notice Plate
WLID	Immediately Dangerous Warning labels/tags
WLAR	At Risk Warning labels/tags
TG5	Tie on Uncommissioned Appliance labels
TG8	Void Property Tag
WL5	Gas Emergency Control Valve labels
WL8	Compartment/Ventilation labels
WL9	Electrical Bonding labels
WL13	Serviced By Label

Gas – Non Domestic

ND1	Essential Gas Safety Non-domestic (Second Edition – Second Revised)
ND2	Commercial Catering and Laundry Non-domestic (Second Edition)
ND3	Commercial Heating Non-domestic (First Edition)

Pocket Guides

USP1	The Gas Industry Unsafe Situations Procedure (Sixth Edition)
CPA2	Combustion performance testing – Non Domestic (First Edition)
VENT1	Boiler Ventilation – Non Domestic (First Edition)

Forms

CP15	Plant Commissioning/Servicing Record (Non-domestic)
CP16	Gas Testing and Purging (Non-domestic)
CP17	Gas Installation Safety Report (Non-domestic)
CP42	Gas Safety Inspection (Commercial Catering Appliances)
CP44	Mobile Catering Vehicle/Trailer Safety Check

Labels

WLID	Immediately Dangerous Warning labels/tags
WLAR	At Risk Warning labels/tags
WL10	Emergency Control Valve labels
WL35	Manual Gas Isolation Valve
WL36	Automatic Gas Isolation Valve

Electrical

Form

CP22 Minor Electrical Installation Works Certificate

Plumbing

Manual Series

HEM1 Hygiene Engineering (First Edition)

Pocket Guides

CDP1 Commissioning of Water Pipework – Domestic (First Edition – Revised)

CWCB Cleansing of Wet central heating Systems (First Edition)

Design Guides

WCH1 Wet Central Heating System Design Guide (out of print)

UVDG Unvented Hot Water Systems Design Guide (out of print)

Forms

CP8 Domestic Unvented Hot Water Storage Vessel Commissioning/Inspection Record

CP20 Central Heating Commissioning/Inspection Record

CP33 Commissioning of Water Pipework

CP34 Central Heating Cleansing Record

CP40 Bathroom Quality Check Sheet

CP41 Combined Pressure Test Record Sheet

Label

TG9 Cleansing service label

Renewables

Manual Series

EEM1 Ground Source Heat Pumps (First Edition)

EEM2 Domestic Solar Hot Water Systems (First Edition)

EEM3 Domestic Biomass Systems (First Edition)

Form

CP11 Solar Thermal Commissioning Record

Labels

WL33 Solar Thermal Installation

WL34 Solar Thermal Pressure Relief Valve

Business

Forms

CP10 Contract of Work & Notice of the Right to Cancel

CP19 Invoice form

Notice

NA1 Sorry We Missed You Cards

NOTES

These pages are intentionally left blank for your use.

These pages are intentionally left blank for your use.

NOTES

These pages are intentionally left blank for your use.

These pages are intentionally left blank for your use.

NOTES

These pages are intentionally left blank for your use.

These pages are intentionally left blank for your use.

NOTES

These pages are intentionally left blank for your use.

These pages are intentionally left blank for your use.